REVEALING ARITHMETIC

Math Concepts from
a Biblical Worldview

KATHERINE [LOOP] HANNON

First printing: 2009

First Master Books revised edition: March 2021

Copyright © 2009, 2010, 2021 by Katherine Hannon. All rights reserved. No part of this book may be used or reproduced in any manner whatsoever without written permission from the publisher, except in the case of brief quotations in articles and reviews. For information write:

Master Books®, P.O. Box 726, Green Forest, AR 72638

Master Books® is a division of the New Leaf Publishing Group, Inc.

ISBN: 978-1-68344-253-0
ISBN: 978-1-61458-776-7 (digital)
Library of Congress Number: 2021934400

Cover by Diana Bogardus

Unless otherwise noted, Scripture quotations are from the King James Version (KJV) of the Bible.

Where marked NIV, Scripture is taken from the New International Version of the Bible®. Copyright © 1973, 1978, 1984 International Bible Society. Used by permission of Zondervan. All rights reserved."

Please consider requesting that a copy of this volume be purchased by your local library system.

Photos by istockphoto.com and GettyImages.com. Image on page 197 from the collection of the Museo National de Anthropología y Arquelogía, Lima, Peru (Photograph by Marcia and Robert Ascher).

Printed in the United States of America

Please visit our website for other great titles:
www.masterbooks.com

For information regarding author interviews, please contact the publicity department at (870) 438-5288.

Master Books®
A Division of New Leaf Publishing Group
www.masterbooks.com

Table of Contents

Acknowledgements 5

Why and How to Use *Revealing Arithmetic* 7

Math Concepts...from a Biblical Worldview? 11

Counting 17

Written Numbers 25

Place Value 31

Comparing Numbers 37

Addition: Foundational Concept 43

Subtraction: Foundational Concept 49

Addition & Subtraction: Multi-Digit Operations 55

Multiplication: Foundational Concept 63

Division: Foundational Concept 71

Multiplication: Multi-Digit Operations 77

Division: Multi-Digit Operations 87

Fractions: Foundational Concept 93

Fractions: Operations 103

Fractions: Factoring, GCF, and LCM/LCD 115

Decimals 125

Percents 135

Ratios and Proportions 139

Types of Numbers (Number Sets) 149

Exponents and Roots 157

Conclusion 165

Math and the Gospel 167

Worksheets 169

Answer Key 187

Appendix A: Mathematicians 193

Appendix B: Different Number Systems 195

Appendix C: Abacuses 199

Appendix D: Math Methods 211

Bibliography 221

Index 225

Acknowledgements

To all those who have kept this project in your prayers — thank you! Your prayer support has played an invaluable part in seeing this project to completion.

To all the moms and dads who have expressed an interest in this project — thank you! I have enjoyed meeting and talking with many of you at conventions. You have been an inspiration to me to keep writing.

A special thanks to the Klein family for their proofing help — you have been a tremendous blessing and encouragement in the final stages of this project. Thank you!

A big thank you to all the people who reviewed this book and provided feedback — your feedback has made a big difference and has helped me sort through what information should be included and how to organize it. I wish to especially thank Deidra Roberson for helping me rethink the layout and for inspiring me with her positive feedback and thought-provoking questions.

I would also like to acknowledge and thank James Nickel, whose book *Mathematics: Is God Silent?* first helped me understand that God was not silent in math.

An extra-special acknowledgment to my family — my dad, mom, and brother — for their support and assistance throughout this project. My mother and brother particularly have often set aside their own pursuits and devoted countless hours to this book, plowing through drafts, offering feedback, and helping me think through content. My brother even took time out of a very busy schedule to review the book, offer technical feedback, and help me completely restructure the format. Thank you, thank you, thank you!

And thank you to the amazing team at Master Books, who saw a vision for this resource and is reprinting it in this new edition, which I hope will make it available to a wider audience.

Above all, I wish to thank the Lord for His faithfulness and unconditional love. He has used this book, including all its challenges and setbacks, to work in my heart and show me how I truly have *nothing* and can do *nothing* without Him.

May the Lord receive glory now and forever.

Katherine [Loop] Hannon

Why and How to Use
Revealing Arithmetic

WHY USE *REVEALING ARITHMETIC*?

Of all subjects, math is probably the one the most people look at as neutral, or "not engaged on either side; not aligned with a political or ideological grouping."[1] Most view math as a set of facts independent from God. Even most Christian math resources, while they add a verse or Bible example to the text, still present math as an independent, neutral fact. Yet this "neutral" view of math is really a humanistic and naturalistic one (see the chapter: **MATH CONCEPTS . . . FROM A BIBLICAL WORLDVIEW?**).

Far from being independent *from* God, math is entirely dependent on God. Math only works *because* it describes consistencies God holds in place. One plus one consistently equals two *because* God faithfully holds this universe together in a consistent fashion. Math testifies to God's faithfulness! If this universe (not to mention our own minds) were the result of random chance, we would have no reason to expect to be able to use math in the first place. Math's overarching message is one of trust and dependency on God. It testifies to God's power, faithfulness, and might.

Sadly, math's message has been twisted underneath a "neutral" guise. Children learn to manipulate numbers on paper. They can tell math works. But they learn to look at math as a system independent *from* God rather than dependent on Him. They adopt an *independent* outlook on math.

This resource is designed to help you "reveal" arithmetic for what it is: a testimony to the Creator. It's easy to lose sight of the big picture when teaching math. This resource will help you remember the purpose of what you're teaching — and pass that on to your children. It will help you answer the "why-are-we-learning-this" questions . . . and show you how to customize your curriculum to show students math's true purpose.

WHO IS *REVEALING ARITHMETIC* FOR?

- **Homeschool parents** looking for a simple guide to help them both understand and teach elementary math concepts (pre-K through about grade 6) from a biblical worldview. You'll find easy-to-read explanations of how the main concepts in arithmetic can be taught from a biblical worldview . . . along with bulleted ideas to use with your children in order to help them grasp the concepts biblically as well.

- **Older students or adults** needing a review of math's basics or an understanding of how math can be seen from a biblical worldview. Math builds on itself — and many

---

1. *Webster's New Collegiate Dictionary*, 1974 ed., s.v. "neutral."

people who struggle with upper-level math do so because they missed something early on. *Revealing Arithmetic* can help older students develop a biblical worldview toward math and build a solid foundation for upper-level concepts.

- **Christian school or co-op teachers** wanting ideas or training on how to modify the curriculum they're using (or will be using) to better convey a biblical worldview to students. The basic principles shared and many of the ideas can be adapted to a classroom setting.

- **College students studying to be elementary school teachers** who want to learn to teach math from a biblical worldview. This books makes a great guide for future elementary school teachers to begin thinking through how biblical principles apply in math.

IS *REVEALING ARITHMETIC* A CURRICULUM?

No. *Revealing Arithmetic* is designed to be used alongside any curriculum. As there is already an abundance of resources focusing on mechanics, you'll find *Revealing Arithmetic* includes just enough of the mechanics to show how the many rules and "facts" in math describe consistencies God created and sustains. It's designed to help you see and teach those mechanics from a biblical worldview rather than to be a curriculum in itself.

HOW IS *REVEALING ARITHMETIC* STRUCTURED?

The book is divided into chapters by concept. The first part of the chapter presents the "big picture" of the concept; the second part offers specific teaching suggestions and ideas. Within the second part, an example curriculum presentation is examined, and suggestions on how to modify it are offered. These examples are included to help you apply the principles discussed. There is material in the appendixes designed to go with some of the ideas given. You'll find there some worksheets that give students an opportunity to see math in action, along with information about historical methods designed to help students see our modern methods as just one way of describing God's creation.

Since this book is designed for parents/teachers of all aged children to use, the ideas often span a wide range of ages and abilities. Please use your own discretion as to whether an idea would be appropriate for your child. A large "H" graphic marks ideas requiring knowledge of more advanced concepts.

Because I personally love history and find it helpful in understanding the worldviews behind a concept, I have frequently incorporated history. If your love is science, you might develop your math lessons with more science.

WHERE DO I START?

Start by reading MATH CONCEPTS . . . FROM A BIBLICAL WORLDVIEW? to get a big-picture overview of how the Bible *does* affect how we view math.

After that, you can **skip around as needed to match what your curriculum is teaching.** If you're covering place value right now with your child, skip to the Place Value chapter and use the background and ideas there to help you teach place value. The chapters are arranged by main concepts to help you quickly find the concept you're teaching, and there's an index in the back.

Note, though, that **you may want to skim any chapters that you skip when jumping ahead to a concept in order to stay in sync with your curriculum**, as the concepts build on each other. The beginning chapters (**COUNTING** through **MULTIPLICATION**) are especially foundational, and many of the principles discussed in them apply to other concepts as well.

If you're using the resource with an older student as a review, have them work through the book in order. The first part of each chapter that explains the concepts from a biblical worldview are going to be what they need to read; you might also have them work some of the problems marked with a big "H."

ECOURSE AND ADDITIONAL MATERIAL

The *Revealing Arithmetic* eCourse (www.MasterBooksAcademy.com) features videos that go along with this material, walking through teaching the different concepts, making an abacus, and more. If you're more visual or auditory, you'll find these videos engaging ways to quickly understand the concept biblically and be equipped to teach it that way too.

The Book Extras section on www.ChristianPerspective.net contains information about additional materials that may be helpful to you as you teach.

Math Concepts...
from a Biblical Worldview?

Math...from a biblical worldview? Those words do not usually appear together. We can see how worldviews matter in history and science, but how can math facts be taught biblically? After all, 1 + 1 = 2 to a Muslim, Christian, and atheist.

For years, if you had asked me how to teach math biblically, I would have suggested adding a scripture verse to the top of the page and including word problems based on biblical data or settings. I later came to realize merely adding a scripture or word problem did nothing to teach the math facts themselves from a biblical worldview. The student still saw math itself as neutral.

But math is *not* neutral, and it *can* be taught from a distinctly Christian worldview! To see how, take a brief look with me at how a Christian worldview applies to science. These same principles apply to math as well.

- **The Bible explains WHERE science originated.** — Genesis 1:1 says, "In the beginning God created the heaven and the earth." God created everything in six days: "For in six days the LORD made heaven and earth, the sea, and all that in them is, and rested the seventh day: wherefore the LORD blessed the sabbath day, and hallowed it" (Exodus 20:11).[2]

- **The Bible tells us WHY science is possible.** — Science is based on observing, repeating, and testing. The whole scientific process presupposes that the universe operates consistently and that man can make intelligent observations. The Bible gives an explanation for both these presuppositions. It tells us a consistent, never-changing God holds all things together (Malachi 3:6; Colossians 1:17) and that this same God created man in His image (Genesis 1:26, 27), capable of subduing the earth (Genesis 1:28). Thus we, unlike animals, can do science!

- **The Bible tells us WHAT to expect as we explore creation.** — We should expect to see both reflections of God's original incredible design, and evidence of destruction and death due to the entrance of sin into the world.[3]

- **The Bible gives us principles to guide HOW we use science.** — God created man to fellowship with Him. Although sin destroyed man's perfect fellowship with God, through Jesus we can again know God and worship Him in all we do. Thus we can worship God as we explore His creation. Since God created us, we are accountable to Him for how we use science.

---

2. The Bible teaches a literal six-day creation and a young earth (about 6,000 years old). For more information, see www.answersingenesis.org.
3. The Bible teaches suffering and death came into the world as result of man's sin (Genesis 3). Before sin entered the world, the universe was "very good" (Genesis 1:31). See www.answersingenesis.org for more resources on understanding death and suffering from a Christian worldview.

Whether we acknowledge it verbally or not, our worldview impacts us continually in science. When we examine a tree, we start with some sort of belief about how the tree got there. Our underlying belief impacts how we view the tree (as God's handiwork, the product of time and chance, etc.). When we see a disease-ridden tree, we either understand the disease as a consequence of man's sin, or else adopt some other explanation (viewing disease as a "natural process," blaming God for the destruction in the world, etc.).

It is easy to see how science can be taught from a distinctly biblical perspective. We can ground ourselves in the Word of God, interpreting the world around us through its lens. Rather than merely learning facts about a tree, we can pause and think of the amazing design before our eyes, letting our hearts turn upward to our Creator. Both Christians and non-Christians have the same facts in science, but interpret them very differently.

In much the same way, worldviews matter in math. While $1 + 1 = 2$ to a Muslim, Christian, and atheist, those three worldviews explain this fact very differently. Join me in a quick overview of the answers the Bible gives to math's where, why, what, and how.

- **The Bible tells us WHERE math originated.** — Since God created everything, He created math too! This does not mean God created the symbols we write, but He created the consistency those symbols represent. Symbols like $1 + 1 = 2$ record the consistent way God causes objects to operate.

- **The Bible tells us WHY math is possible.** — Why does $1 + 1$ consistently equal 2? Because God both created this consistency and keeps it in place! Math's very ability to work is a testimony to God's faithfulness. If this universe were simply a random collection of chemicals, or if we served an inconsistent God, we would have no reason to expect objects to add or subtract consistently.[4]

    In the Book of Jeremiah, God pointed out to His people the consistent way He has chosen to govern the universe, reminding them He would keep His covenant with them. It is as if math were shouting out at us, "You can trust God!"

---

4. While many mathematicians have tried to explain math's ability to work without acknowledging God, sooner or later something has come along to dislodge their theory.

    You may wish to look at a presuppositional book on logic, such as *The Ultimate Proof* by Dr. Jason Lisle. In this book, Dr. Lisle makes the case that only the biblical worldview makes sense of the laws of logic and that, apart from the biblical God, there is no way to really prove or know anything. He also makes the case that it is the biblical God, not any other God, who makes sense out of logic. (I would add that this is true of math too. Only the biblical God has the characteristics necessary to account for what we see in logic and math.)

"This is what the Lord says: 'If I have not established my covenant with day and night and the fixed laws of heaven and earth, then I will reject the descendants of Jacob and David my servant.'"

JEREMIAH 33:25-26A (NIV)

In order for math to work, there not only have to be consistencies throughout creation, but we have to be able to recognize those consistencies! Again, the Bible gives us the framework for understanding this. It teaches us God created man in His image and gave him dominion over the earth (Genesis 1:27–28). Hence we should expect to be able to develop ways to record the consistencies God placed around us.

We should also expect our thinking to, in a very limited way, take after our Creator. Many purely intellectual mathematical theories end up corresponding with reality because *our minds were created by the same Creator who created all things.*

- **The Bible tells us WHAT to expect as we use math.** — Although we tend to confine math to a textbook, it actually goes hand-in-hand with science! Since math records real-life consistencies, we should expect it to help us explore God's creation, showing us both reflections of God's original incredible design and evidence of destruction and death due to the entrance of sin into the world. We should also expect to be able to use math to help us in the various tasks God has given us to do, be they around the house or on the job.

- **The Bible gives us principles to guide HOW we use math.** — Through Jesus, we can again know God and worship Him as we use math! Since God created us, we are accountable to Him for how we use the gift of math He has given us.

So how can math be taught biblically? In much the same way science can! Just as we would worship God while studying a tree, recognizing Him as the Creator and Sustainer, we can worship God as we study math concepts, recognizing He created and sustains the consistency the concept represents.

The problem is, nearly all math textbooks approach math as a neutral concept. We typically look at math as facts to solve on paper — facts that have always been there and that we can never question. This "neutral" approach is not neutral at all — we have to give the credit for math's ability to work somewhere. When we do not give the credit to God, we end up buying into a naturalistic and humanistic view on math.

- **Naturalism** — Most math books present math as a self-existent fact. For example, one algebra book, when talking about properties says we "are stuck with properties because they are what they are."[5] Yet properties represent consistencies all around creation. Calling properties or other math facts self-existent subtly calls the consistencies throughout creation self-existent as well, which is a naturalistic and evolutionary idea!

- **Humanism** — A student studying math might also get the impression man created math. After all, man *did* develop many of the tools and techniques we use in math. But man did *not* develop math! Regardless of when man discovered them, math principles

---

5. Saxon, *Algebra 2*, 8.

have been in action all around us since God created the world. When we give man the credit for math's ability to work, we buy into humanistic thinking and begin to place our faith in man's intellect instead of in God's Word.

There is a fundamental worldview conflict going on in math, as in science and every other subject. If we want to see God in math, we have to do more than add a scripture verse or biblical example to a presentation. We have to completely change our approach to math.

How? Throughout this book, you will notice four main tactics used to "reveal" math concepts.

1. **Rewording the presentation.** Often, textbooks have students memorize rules without really understanding why the rules work. This robs the student of seeing how the rules describe a real-life principle God created and sustains and leaves him looking at math as man-made. Thus you will notice I have tried to explain *why* various rules work. The goal is to show how each rule, far from being self-existent or man-made, describes a real-life consistency God created and sustains.

2. **Sharing the history of the concept.** In recent years, math has been predominately isolated from history. Yet seeing the history of a concept can go a long way in helping students see it from a biblical worldview! When students only learn one method or symbol, it is easy for them to mistake that method or symbol for math itself, thereby leaving them thinking of math as a man-made system (after all, man did develop the method and the symbol). History, however, reveals many other methods and symbols, thereby helping us see each method or symbol as but *one* way of describing something God created and sustains.

3. **Applying the concept in a real-life situation.** Can you imagine spending years reading instruction manuals on how to work a sewing machine without ever touching one? How silly! Is it not equally silly for students to spend years of their lives studying math concepts without being shown how to actually apply their knowledge outside of a textbook? Math has been isolated to a textbook in a large sense because of a false worldview. Apart from God, it does not make sense why math works, nor do we have the same motivation to use it as a useful tool to serve God. For years, math has been taught as an intellectual pursuit — as something to show how smart we are or to get us into a university.

   God, however, urges us to do everything we do as unto Him. Our goal should not be to puff ourselves up with our ability or impress others, but to acquire a skill we can use in the tasks God has given us, all the while praising Him!

   I have included suggestions on how you can let your child apply math concepts in his own life. These suggestions should both help your child connect math with a way of recording real-life consistencies and help him learn to use math as a useful skill rather than empty head knowledge.

4. **Using the concept to explore an aspect of creation.** Math is the tool scientists use to discover the intricate way God created different aspects of creation. For example, math helps us calculate the distance to stars, the special pattern God put within a sunflower, the order within our own bodies, and much, much more. Throughout the book, I have interspersed a few examples of math's use in science. My goal in these examples is to help students see math outside a textbook and understand how it aids in exploring God's creation, giving us glimpses of our Creator at every turn.

Much as our heart pumps blood through our entire bodies, our worldview gives life to everything. If we approach math from a dependent worldview — a worldview acknowledging math's complete dependence on God and inability to exist without Him — it gives a whole new life and color to every aspect of math. Math ceases to be a paper exercise and becomes a real-life tool we can use in the tasks God has given us, whatever they might be. Math connects with history and science, taking its place, not as a mysterious intellectual pursuit, but as yet another area in which we can worship our Maker and see reflections of His character.

Best of all, math becomes an encouragement to us in our Christian walks. For as we see God's faithfulness in holding this universe together, it encourages our hearts and reminds us to live in trust instead of fear or pride.

I hope by now you are as excited as I am at digging into math concepts. I invite you to begin "revealing arithmetic" with me.

Counting

Most of us learned to count at a very young age. We proudly exclaimed to everyone who would listen, "Look at me! I can count to ten! One, two, three, four...." Although simple enough for a young child to learn, counting forms the basis for all other math concepts, as shown.

At first glance, counting appears neutral. After all, how could counting "one, two, three, four..." have anything to do with God or a worldview? Although counting seems neutral on the surface, digging a little deeper reveals a different story.

EXPONENTS: *Repeated multiplication*	EXTRACTION OF ROOTS: *Repeated division*
MULTIPLICATION: *Repeated addition*	DIVISION: *Repeated subtraction*
ADDITION: *Counting forwards*	SUBTRACTION: *Counting backwards*
COUNTING	

The first few chapters of Genesis give us a foundation for building a biblical worldview toward every aspect of life, including counting. Take a look with me at different events in Genesis 1 through 3 and what they reveal about math in general, and counting in particular.

God Created All Things

The Bible starts by telling us God created the heavens and the earth.

> *In the beginning God created the heaven and the earth.*
>
> GENESIS 1:1 (KJV)

God is the creator; He created *all* things.

> *For by him were all things created, that are in heaven, and that are in earth, visible and invisible, whether they be thrones, or dominions, or principalities, or powers: all things were created by him, and for him:*
>
> COLOSSIANS 1:16 (KJV)

Notice it does not say God created all things with the exception of math. It says He created *all* things. He created math!

This does not mean God created the symbols on the piece of paper we have come to associate with math. Rather, He created the way real-life quantities interact in this world. Man developed the symbols to represent these real-life consistencies God created and sustains. Math records aspects of God's creation!

When we count, then, we are using names to describe quantities God created.

God Created Man in His Image

In Genesis 1:27, we learn God created man as the crowning jewel of His creation. Man, unlike animals and plants, was made in God's image (Genesis 1:27), capable of fellowship with Him.

> *So God created man in his own image, In the image of God created he him; male and female created he them.*
>
> GENESIS 1:27 (KJV)

How does this truth apply to math? It both 1) explains why we can count and 2) holds us accountable to God, our Creator, for how we use the gift of counting. Man, unlike the animals,[6] can count because God created us in His image, capable of seeing and classifying the order He placed around us. Our very ability to count is a God-given gift for which we are accountable to God!

God Gave Man Work to Do

Genesis 1:28 tells us God gave man authority over the rest of creation and the task of subduing the earth.

> *And God blessed them, and God said unto them, Be fruitful, and multiply, and replenish the earth, and subdue it: and have dominion over the fish of the sea, and over the fowl of the air, and over every living thing that moveth upon the earth.*
>
> GENESIS 1:28 (KJV)

God put Adam and Eve in the Garden of Eden to tend it. He gave man work to do.

> *And the LORD God took the man, and put him into the garden of Eden to dress it and to keep it.*
>
> GENESIS 2:15 (KJV)

God did not give man tasks without giving us the ability to do them! He created us capable of doing all He gave us to do. He created us capable of observing and naming creation. In counting, we observe and name quantities. The concept of counting is part of God's provision to help us complete the tasks He gives us.

We use counting all the time. When you set the table, you count the correct number of plates and silverware. When you fold the wash, you count as you grab two socks and fold them together (you just counted two!). When you take medicine, you often count the correct number of teaspoons or pills. When you cook, you count. Perhaps you count two cups of flour, or put two cookies in a lunch pail. Aren't you glad God gave us the ability to count?

---

6. Although some animals can be trained to mindlessly repeat numbers, they do not have the understanding and ability to really count and explore the consistencies throughout creation using math like humans.

Adam Named the Animals

On the very first day of mankind's existence, God brought the animals to Adam, and Adam named them.

> *And out of the ground the LORD God formed every beast of the field, and every fowl of the air; and brought them unto Adam to see what he would call them: and whatsoever Adam called every living creature, that was the name thereof.*
>
> GENESIS 2:19 (KJV)

The naming process bears many similarities to the counting process. In naming the animals, Adam 1) observed God's creation (the animals) and 2) assigned names to describe the different animals. In counting, we 1) observe God's creation (the quantities around us), and 2) assign names to different quantities. "One" is the name we use in English to describe a single unit — a single pen, dollar, toy, etc. "Two" is the name for a group of two units of anything.

Thus right in the Garden of Eden, we have an example of man using his God-given ability to observe and name to complete the task God had given him! Notice God brought the animals to Adam for naming — Adam was in God's presence while observing and naming. Prior to the fall of man, both God and man engaged in the naming process together.

God created all things "very good" (Genesis 1:31). No sin, suffering, or death marred the world like today. Adam's ability to explore God's creation, as well as to walk and talk with God, was originally perfect.

As we continue reading in Genesis, however, we learn about an event that changed the world forever: the fall of man.

Sin Ruined an Originally Perfect Creation

Genesis 3 tells us about the entrance of sin into the world. Man's sin, or rebellion against God, changed everything, including counting. Our ability to classify and explore God's creation is no longer perfect. We now see but darkly (1 Corinthians 13:12).

The whole creation now "groaneth and travaileth in pain" (Romans 8:22). Death came into the world as a result of sin (Genesis 2:17, 1 Corinthians 15:22). As we count, we should expect to see evidences of both God's original design and the fall of man. And we do!

For example, if we use counting to explore our hands, we find we have four fingers and one thumb on each hand. Counting helps us see the design God placed within hands! At the same time, though, if we were to count the fingers on every person's hands, every once in a while we would find a person with a finger cut off, evidence we live in a fallen universe.

Through Jesus We Can Again Fellowship with God

Thankfully, God's message to man does not stop with man's sin. In the middle of cursing the serpent, God proclaimed this hope: an offspring of the woman would one day crush Satan's head forever.

> *"And I will put enmity between you and the woman, and between your offspring and hers; he will crush your head, and you will strike his heel."*
>
> GENESIS 3:15 (NIV)

Before man's sin, God had clearly told Adam, "But of the tree of the knowledge of good and evil, thou shalt not eat of it: for in the day that thou eatest thereof thou shalt surely die" (Genesis 2:17). Adam knew the penalty for sin was death. God keeps His Word. As soon as Adam sinned, death entered the world.

Man's sin, however, had not caught God by surprise. Right there in the Garden of Eden, God did something to foreshadow what He would one day do to take the penalty for sin and redeem those who would believe Him. God killed an animal and clothed Adam and Eve.

While the blood of an animal could never take away man's sin, it foreshadowed the blood God's own Son, Jesus Christ, would one day shed to save man from eternal death and destruction. Through Christ Jesus, our relationship with God can be restored. We can enjoy eternal life rather than eternal death and torment *because* Jesus paid the penalty for sin for all mankind.

What does all this mean for counting? Through Christ Jesus we can again worship and fellowship with God while we count! In fact, Colossians 3:17 (NIV) tells us, "And whatever you do, whether in word or deed, do it all in the name of the Lord Jesus, giving thanks to God the Father through him." Notice it does not say "whatever except math." It says *whatever*. As we count and use math, we have the opportunity of worshiping God and depending on Him!

Conclusion

While we could explore many other scriptures relating to our view of counting, hopefully this short overview sufficed to illustrate that counting is *not* neutral. Already we have seen

- Where counting came from (a gift from God) and what it is (a way of describing His creation)

- Why we see both order and problems as we use counting (sin ruined God's originally perfect creation)

- How we can use counting (in God's presence, knowing we are accountable to Him)

The principles we have discussed in this chapter give us a foundation, not only for counting, but for the rest of math as well. Because God created all things and made man in His image, we are able to use math. Every math concept rests on Him.

TEACHING SUGGESTIONS AND IDEAS

Objective: *To lay a biblical foundation, connecting math in your child's mind with a God-given ability to record and explore the quantities God placed around us.*

Specific Points to Communicate:

- *Names like "one" and "two" are ways of describing real-life quantities.*
- *Counting is a gift God gave man for which man is accountable to God.*

You can teach your child to count in much the same way you taught him to speak and identify colors — as a part of normal, everyday life. Keep your eyes open for opportunities to have your child count — toys, food, crayons, and other household objects all make excellent math manipulatives.

For example, if your child needs to put away a pile of blocks, have him count each block as he puts it back. When he is finished, point out he used words like *one* and *two* to help him name quantities (the quantity of blocks). Explain that *one* is the word we use to name a single quantity — be it a block, plate, toy, pencil, apple, or idea.

Having your child count real-life objects serves an important purpose. It teaches him to view numbers as ways of describing real quantities. The next step is to help him understand God gave us this ability to count and record quantities.

You can teach your child this by simple comments throughout the day like, "Very good! You were able to count those apples! Do you know why? Because God gave man the ability to count and explore." At some point, read Genesis 2:19–20 with your child, mentioning how right from the very beginning, man has been naming God's creation. On the very first day of Adam's life, God brought all the animals to him so he could name them.

Counting, along with the rest of math, is one way man names and explores God's creation. Adam used names to describe the animals, and we use numbers to describe quantities. Adam was fellowshipping with God while he named the animals, and we can fellowship with God while we use numbers because — and only because — Christ died on the cross for us.

It is important to note that numbers describe quantities, not specific objects. You can show your child this by having him count a variety of different objects, highlighting how we describe "three" the same way, no matter what type of object it might be. The name describes the quantity, not the item.

Example

For the other concepts in this book, we will take a look at a typical textbook presentation and discuss ways to change it in order to convey the above objective and specific points to communicate. But since most children learn to count long before they ever open a math book, I felt looking at a textbook example for counting would be more confusing than helpful.

Ideas

◆ **Look for simple opportunities to let your child use counting to explore God's creation.**

» **Our Bodies** — For a wonderful — and simple — opportunity to let your child explore God's universe with counting look no further than your own body. Have your child count the fingers on your hand and on one of his hands. Both hands have the same number of fingers. Talk about the order God placed in hands, as well as the effects of sin we see (sometimes people have a finger cut off). Note: You do not have to stop with hands! You could have your child count eyes, ears, noses, toes, joints, etc.

» **Stars and Hairs on Our Heads** — Using counting to explore the stars and hairs on our heads points us both to God's greatness and His personal care. Begin by heading outside and asking your child to count the stars. He should quickly see the impossibility of ever counting the stars, especially once you mention the existence of galaxies not even visible with the naked eye. While we cannot even count the stars, God knows their number, calls each star by name, and keeps each one in its place! What a mighty, powerful God!

> *Lift up your eyes on high, and behold who hath created these things, that bringeth out their host by number: he calleth them all by names by the greatness of his might, for that he is strong in power; not one faileth.*
>
> ISAIAH 40:26 (KJV)

> *He telleth the number of the stars; he calleth them all by their names.*
>
> PSALM 147:4 (KJV)

Next have your child try to count the hairs on his head. After he has tried and failed, share with him Matthew 10:29-31.

> *Are not two sparrows sold for a farthing? and one of them shall not fall on the ground without your Father. But the very hairs of your head are all numbered. Fear ye not therefore, ye are of more value than many sparrows.*
>
> MATTHEW 10:29-31 (KJV)

By showing us how unable we are to even identify all the stars and hairs on our own heads, counting helps us better comprehend how much greater God is than us! Yet the God whose power knows no bounds and who knows the number of the stars cares personally for each one of us — He has even numbered the hairs on our heads.

» **Other Aspects of Creation** — Other aspects of creation you could have your child explore with counting would include the number of petals on a flower, the number of tomatoes on your tomato plant, or the number of leaves on a vine. You could even check out a book from the library on plants and look at how counting the number of leaves helps us identify some plants (such as poison ivy). Counting various objects in God's creation can provide a lot of fun as well as opportunities to marvel at God's greatness and help your child view math as a useful way of exploring God's creation!

- ◆ **Reinforce counting through real-life examples.** Use the times you find yourself counting to reinforce counting's usefulness. You could make a game out of finding real-life uses for counting — see who can notice the most!

  As your child begins to see counting as practical, pause together to thank God for creating us able to count and for redeeming us through Christ Jesus so we can know God and delight in Him while we count.

- ◆ **Play games involving counting.** Since many games involve counting (and children typically love games!), why not play a game with your child? Any board game requiring counting spaces will work. Or you could "make up" your own game. For example, you could play "Go Get" by asking your child to get a certain number of items, or play "Mother May I" by asking your child to take a certain number of steps forward or backward. Math allows us to name quantities, such as the number of spaces to move on a board game, the number of items to get, or the number of steps to take. Even while playing a game, we use math to help express a quantity in a similar way to how Adam used names to express an animal.

- ◆ **Have your child count objects in books.** As you read picture books, have your child count different objects or animals on the page. At times, you might remind them counting helps us describe the quantities around us.

PARTING NOTE

You may find it helpful to create an idea notebook to aid you while you teach. Use this notebook to jot down all the practical uses for math you discover. If you find yourself using math to measure fabric for a dress, write that down in your notebook. If you read about a new scientific discovery in the newspaper in which math was applied, make a note of it. Then on those days when you need ideas, you can use your notebook for inspiration.

Written Numbers

We use written numbers so often it is hard to imagine life without them. But what are written numbers? And how did they get here?

The same biblical principles we examined in the last chapter apply to written numbers. Much as the words *one*, *two*, and *three* describe, or name, quantities in God's creation, the symbols *1*, *2*, and *3* represent those quantities on paper. The symbols we use (*1*, *2*, *3*, etc.) in math, as well as the words (*one*, *two*, *three*, etc.), are part of *one* language system to describe real-life quantities and consistencies God created and sustains.

I emphasized the word *one* because we tend to think of our modern numerals as *the* way to record quantities. Nothing could be further from the truth!

All throughout history, men have used different words and symbols to describe quantities. Taking a look at these different methods aids in viewing our current method as a language rather than as an absolute structure. This in turn helps us see math as a description of God's creation, not some sort of man-made absolute. God created quantities and gave man the ability to find methods to observe and record them.

To better illustrate this point, let us head back to Genesis again and take a look at some early uses of numbers, as well as at a key event that forever changed how we communicate about quantities. Because counting and written numbers are so closely intertwined (one is an oral way and the other a written way of communicating about quantities), we will be looking at both.

Back to the Beginning

Right from the beginning of history, we find men using numbers to communicate and help them with tasks. The Bible tells us Cain built a city (Genesis 4:17). Although we do not know exactly what type of city he built or what process he used, we have no reason to suppose Cain did not use math and numbers. It would make sense for Cain to have used some sort of measuring device or number system while building his city.

We do know Cain's great, great great, grandson, Lamech, spoke of numbers, saying, "If Cain shall be avenged sevenfold, truly Lamech seventy and sevenfold" (Genesis 4:24). We also know God used numbers to communicate to Noah instructions about building the Ark. He told Noah, "The length of the ark shall be three hundred cubits, the breadth of it fifty cubits, and the height of it thirty cubits" (Genesis 6:15b).

After the Flood, an event happened that had massive effects on numbers: the Tower of Babel. Prior to this event, "the whole earth was of one language, and of one speech" (Genesis 11:1). Thus, men would have used the same words to describe quantities.

At the Tower of Babel, men misused the ability God had given them to communicate and sought to unite against God and try to make a name for themselves. The project stopped abruptly when God came down and confused their languages, thereby scattering them across the earth.

The Tower of Babel accounts for the many different language systems we find, including the different words used to describe numbers. As men spread out across the earth and became unique cultures, we would expect them to use unique mathematical symbols as well.

Figure 1 shows some of the many symbols cultures have used to express a single quantity (what in English we would call "one").

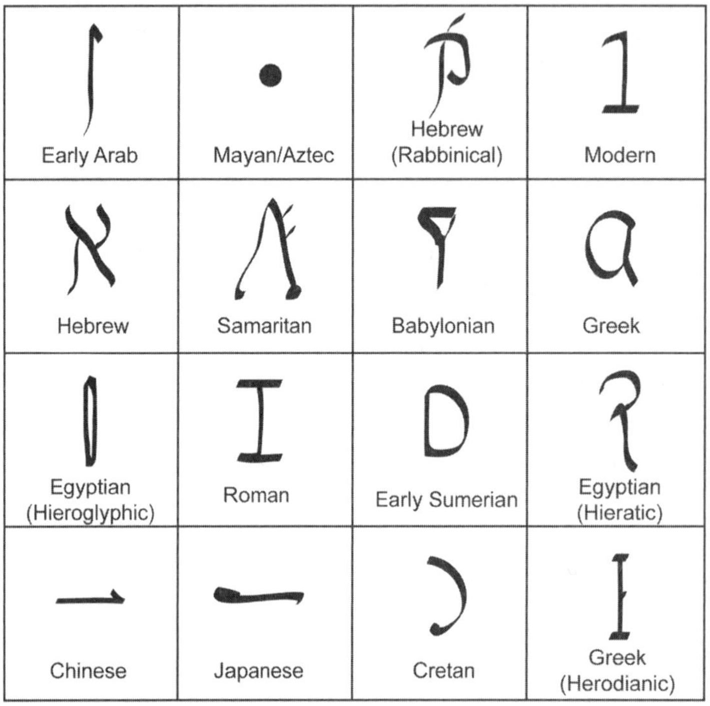

Early Arab	Mayan/Aztec	Hebrew (Rabbinical)	Modern
Hebrew	Samaritan	Babylonian	Greek
Egyptian (Hieroglyphic)	Roman	Early Sumerian	Egyptian (Hieratic)
Chinese	Japanese	Cretan	Greek (Herodianic)

Figure 1: Different Symbols to Express a Single Quantity

There is another important lesson to take from the Tower of Babel. God gave us the ability to communicate for a purpose — and that purpose was not to build a name for ourselves. As we learn to use written numbers, we will want to remember our purpose is not solely to learn something and impress others with our knowledge. We want to worship God in math and use it to accomplish tasks He puts before us, all the while depending on and trusting Him.

Math and Science Go Hand in Hand

Because written numbers represent real-life quantities, we can use them to help us explore the universe around us. Although we tend to separate math and science, they go hand in hand. Math is the tool scientists use to learn about the universe.

For example, written numbers helped Danish astronomer Ole Christenson Rømer (1644-1710)[7] realize light had a definite speed. Written numbers were used to record when Io (one of Jupiter's moons) became visible after an ellipse. Over the years, a pattern began to emerge. During certain times of year when the earth is farther away from Jupiter, it took longer for Io to appear after an ellipse. The question was...why?

The answer...because the light takes longer to reach the earth when the earth is farther away! By recording times or looking at times others had recorded, then using more math and written numbers, Rømer realized light had a definite speed. Rømer (or another astronomer — sources disagree about who actually made the estimate)[8] estimated the speed of light to be 214,000 kilometers per second. This estimate, while not completely accurate (the actual speed is closer to 300,000), was amazingly close.[9] Recording data using written numbers helped Rømer realize light travels at the same speed all the time!

As we use math to explore creation, we also see glimpses of God's character throughout His creation. For example, light's consistent speed testifies to God's consistency. Hebrews 1:3 tells us Jesus is "upholding all things by the word of his power." He is the One who holds light together and causes it to operate in the same predictable fashion! The speed of light is constant because God is constant. He does not change based on our moods or the situations in which we find ourselves. Light's consistent speed reminds us God is the same today as He was when He parted the Red Sea and turned water into wine. What an encouragement to walk forward in confidence knowing our circumstances or feelings have not changed God!

For I am the LORD, I change not; therefore ye sons of Jacob are not consumed.

MALACHI 3:6 (KJV)

This one example illustrates how written numbers help us explore God's creation, in the process turning our eyes upward to God's character.

Conclusion

Our current method of writing numbers is one way to express real-life quantities on paper. Right from the beginning of history, men have been using numbers to communicate, help them in tasks, and explore creation. As we use written numbers to learn about the universe, we have opportunities to see glimpses of God's character and worship Him.

TEACHING SUGGESTIONS AND IDEAS

Objective: *To present written numbers as ways of recording on paper the quantities we find around us.*

---

7. Also spelled Roemer. Note: This story is too complex to share with young children, but is provided here for your reference and background information.
8. The sources I consulted disagreed as to whether Rømer actually gave a guess as to the speed of light or whether someone else made the estimate based on his writings. Either way, written numbers played an important part.
9. Wilson, *Astronomy Through the Ages*, 96.

Specific Points to Communicate:

- *Number names and symbols express real-life quantities; after the Tower of Babel, many different names and symbols arose.*
- *Written numbers are a useful tool we use all the time.*
- *We can praise and worship God as we learn about and use numbers.*

You can teach written numbers much as you did counting — through the use of real-life manipulatives (toys, food, crayons, or other household objects). Instead of teaching your child how to name the quantity with *words*, you are now teaching him the written *symbol* used to "name" the quantity.

Part of learning to write numbers is learning the mechanics and motor skills of forming the symbols 0-9 on paper. As you teach the mechanics of writing symbols, do not lose sight of what those symbols represent! Try to frequently have your child look at a small group of objects (toys, food, crayons, or other household objects) and write a number to represent them. Remind him the symbol he put on paper is just one way of naming a certain quantity! As you did with counting, continue to "connect the dots," reminding your child why he can learn this system for recording quantities — because God created him with this ability and created a universe that can be counted.

Example

Below is an example of the type worksheets most curriculums have children fill out when learning to write numbers.[10]

While there is nothing wrong with the above worksheet and it could be very helpful, notice how adding something like the following helps the student develop a biblical view of math.

> *How many* ___ *[toys, crayons, etc.] do I have here? That is right, we would say I have two* ___. *"Two" is the name we use to describe this quantity.*
>
> *If I wanted to write down on paper how many* ___ *[toys, crayons, etc.] I had, can you think of how I could do that?* [Pause, listen, and respond to the child.]

---

10. A Beka Book, *A Beka Numbers Skills K,* 9.

It is not easy to write out a whole word every time we want to represent a quantity. Different people groups have developed different symbols. In our country today, we represent two like this: 2.

Here is a sheet you can use to practice drawing this symbol. Since you are learning to write 2, see how many 2s you can find written in different places today. I will try to find as many 2s as I can too, and we will see who can find the most!

Notice how these tiny additions help the child see math as a language system to describe quantities God created and sustains. The next section shares some other simple ways you could present or reinforce a biblical perspective on written numbers.

Ideas

- **Show your child how God used written numbers to communicate His message to man.** All throughout the Bible, we find written numbers — look at the genealogies for a whole host of them! The numbers in the genealogies serve an important purpose — they trace the lineage of God's promised Redeemer, Jesus, showing us He did indeed come into history exactly as God had promised.

- **Integrate written numbers into everyday life.** To help your child connect the symbols he has learned to write with a way of recording quantities, why not have him count and write down the quantities of various real-life objects? We use written numbers all the time, so finding ideas should not be difficult. (There are numbers on clocks, cereal boxes, road signs, street addresses, pages of books, telephones, etc.). There are two worksheets in the worksheet section (page 169) to help you get started.

- **Teach your child how to dial important telephone numbers (like 911 and his dad's work number), address envelopes, use a calendar, and tell time.** We use numbers to express more than quantities. Numbers help us label telephone numbers, addresses, etc. When you teach telling time, take a look at Genesis 1. There you will find the basis for different time dividers (the day, week, year, and seasons).

- **Have your child apply written numbers by measuring various objects around the house and recording his answers.** You can have lots of fun with this! What child does not want to know his height?

- **Explore how other cultures wrote numbers.** Some children might enjoy learning how to write quantities in different number systems. Besides providing some entertainment, learning other number systems can help your child think about math as one way of describing God's creation rather than as some sort of system that has always been there. You could even integrate math with history by looking at the number systems of the different civilizations you study! Of course, you will want to use your own judgement as to whether looking at other systems would confuse your child. Appendix B explains several past number systems. If your child has not learned place value yet, you will want to stick with the ones in the

"Fixed Value Systems" section. Reminder: The [H] means this idea is for older students reviewing the concept.

PARTING NOTE

Right from the beginning, start building a biblical understanding of math! Throughout the other concepts we examine, we will be building on the foundations established here. All of math is a way of observing and recording the quantities and consistencies around us.

Place Value

Can you imagine how confusing it would be if we had a different symbol for every single number imaginable? Or if, in order to write forty, we had to draw forty symbols, or even four symbols? Math would be much more time consuming and could become quite difficult!

Fortunately, we use something called place value to help us. In a place value system, the location of a symbol affects its value. For example, in the Hindu-Arabic place value system we typically use, 1 represents one if it is by itself, but if 1 is followed by another number, it represents ten. If 1 is followed by two numbers, it represents one hundred. Utilizing place value, we can use ten symbols (called digits) — 0, 1, 2, 3, 4, 5, 6, 7, 8, 9 — to represent any quantity.

Historic Number Systems

Our Hindu-Arabic place value system is not the only system for recording large quantities, nor is it always the best system to use for every situation. This place value system is only *one* way men have agreed to record quantities. Different people groups and cultures have used and developed different systems for recording and working with numbers.

Some cultures, like the Egyptians, did not use a place value system, except sometimes when dealing with extremely large numbers.[11] To represent ninety one, the Egyptians used their symbol for one and then repeated their symbol for ten nine times, as shown.

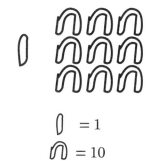

Other cultures, like the Babylonians, used a place value system based on sixty instead of on ten. Still other cultures chose to use other methods besides writing to record quantities. The Incas recorded quantities by tying knots on a device called a quipu (kē′pōō).[12]

Today, we use what is called a Hindu-Arabic system, so named because it comes from the Hindu system which the Arabs adopted and brought to Europe.[13]

Note: For more details on different number systems, see Appendix B.

The history of number systems shows us math is not man-made. Think about it. All these different cultures developed *different* ways to record and work with quantities, but, although their *methods* varied, the underlying *quantities* and *consistencies* remained the

---

11. Toward 1600–2000 B.C., the Egyptians began to use something similar to a place value system to represent extremely large numbers. "In the oldest example hitherto known, the symbols for 120, placed before a lotus plant [symbol for 1,000], signify 120,000. A smaller number written before or below or above a symbol representing a larger unit designated multiplication of the larger by the smaller." Cajori, *History of Mathematical Notations*, 1:14.
12. Pronunciation from *The American Heritage Dictionary of the English Language*, 1980 New College Edition, s.v. "quipu."
13. Groza, *Survey of Mathematics*, 52.

same. In order for a number system to work, one plus one *had* to equal two, not three. There has to be something beyond symbols in math.

The Bible tells us God created and sustains all things and made man in His image (Colossians 1:16–17; Genesis 1:27). These two truths account for math's existence. They explain the underlying consistency different number systems describe, as well as man's ability to recognize this consistency.

Our place value system itself, as well as other number systems, would be absolutely pointless if only man-made. These systems prove useful *because* they describe something God created and sustains! Even our place value system depends on God — numbers like 40 would have no meaning if objects did not count consistently, making 4 groups of 10 consistently equal the same number.

The history of numbers helps us break away from viewing math as a body of self-existent, eternal truths. Seeing different systems reminds us each one is but a tool — a device — to record quantities.

It is also interesting to note how much different civilizations accomplished using math. The Egyptians discovered and used fractions, negative numbers, algebra, and advanced geometry as they ruled an empire and constructed incredible pyramids. The Babylonians used irrational and negative numbers, and laid the basis for the time system we use today. The Mayans implemented the concept of zero and kept detailed and extremely accurate calendars. The Incas not only operated a huge empire spanning more than 15,000 miles, but baffled their Spanish conquerors by their ability to record the tiniest details as well as the largest ones on their quipus.

Although many modern books give the impression past civilizations were inferior to us, this simply is not the case! God created man completely formed and fully intelligent. Right from the beginning, we should expect to find men seeking to understand the world around them and developing tools to help them in this process. This is exactly what we find.

All throughout history, men have been using the ability God gave them to record quantities. Our place value system is one useful way of describing quantities and consistencies He created and sustains!

Place Value — A Useful Tool

Place value simplifies tasks and proves quite useful. To see how, try adding Roman numerals on paper — it is not easy! Place value allows us to use algorithms, or written methods, to work with numbers.

Much as different tools prove useful for different tasks, different place value systems prove useful in different areas. Electronic devices operate by using a binary place value system (a system based on 2). Since it is based on 2, the binary system translates easily into "on" and "off" flows of electricity. In this system, forty would be written 101000. Computer programmers also use a place value system based on 16 called the hexadecimal system — each of these systems proves useful in a different way.

The binary, hexadecimal, Hindu-Arabic decimal, and other number systems are all useful in different situations. Each one is a useful tool to help us explore God's creation and complete the tasks He has given us to do! And who knows what other number systems we might one day find handy?

In Conclusion

All the place value systems in the world would be meaningless and useless unless they helped us record something real and consistent in the world around us. Place value systems only prove valuable because they represent God's creation.

Each place value system helps us communicate and express quantities. We can only develop place value systems because God created us capable of exploring and recording His creation — and because He created and sustains a consistent universe.

TEACHING SUGGESTIONS AND IDEAS

Objective: *To help your child see each new number as a way of recording on paper the quantities God created and placed around us, understand God gave man the ability to develop methods for writing quantities, and see place value as one of those methods.*

Specific Points to Communicate:
- *An understanding of how our place value system describes quantities.*
- *A correct perspective of each number system as one system to express quantities, not as some sort of self-existent fact in itself. (This reinforces math as dependent on God rather than independent from Him.)*

Before you begin teaching place value (the concept of letting a digit's place determine its value and, more specifically, the place values in our current number system), make sure your child is comfortable grouping objects together. Have him practice grouping household items into groups — particularly groups of 10.

As you present place value, continue to "connect the dots." Our place value system of writing numbers is one out of many methods by which we can represent quantities. Place value will be continually reinforced in other concepts. At this point, it is not necessary for your child to understand everything — the basics will do!

Example

While the methods of presenting place value vary greatly, most curriculums demonstrate place value using some sort of object or manipulative. Consider the following example.

> *The number 35 has two places. The 3 is in the tens' place. The 5 is in the ones' place.*[14]

The presentation proceeds to give very helpful pictures illustrating with cube manipulatives how 35 represents 3 groups of 10 and 5 single blocks.

---

14. Cummins, *Horizons Mathematics 1: Teacher Handbook*, Part 2, Lesson 5.

Notice this presentation *does* demonstrate how place value works (with the manipulatives), but *does not* present place value as only *one* way of describing quantities God created and sustains.

Students are left to draw their own conclusions about how place value got here. This vagueness often leads to confusion and, worse still, to students thinking of math as man-made or self-existent rather than as a way of describing what God created.

By introducing place value as *one* way of describing quantities, you can help your child see place value differently. Below is one example of how you could do this.

> *There are lots of different ways to record numbers! Today, we are going to learn about one useful way. In this way, we think of quantities in terms of groups. We could look at the number thirty five as 3 groups of ten and 5 ones.* [Have the child count out 35 manipulatives or objects and form them into 3 groups of ten and another group of five.] *One way to represent that on paper is like this: 35. When a number is written to the left of another number, we will have it represent groups of 10. That's how we know the 3 we wrote in 35 stands for 3 groups of 10 and not just 3 individual objects. We will call it the ten's place...*

Or you could start by having the child form numbers on an abacus (see the "Ideas" section), then gradually lead him to discovering the method for recording place value we use today. This not only leaves him with a clear understanding of our modern notation as *one* way of describing quantities, but also teaches him to begin using his God-given ability to think and reason for himself instead of teaching him to rotely memorize a method.

Depending on the extent to which your curriculum demonstrates how place value works, you might also want to add more manipulatives or applications of place value. Connecting place value with real life is an important aspect of seeing place value as a tool to represent God's creation. You can easily do this by having your child group items, such as beans or raisins, into groups of ten and practice recording them using place value.

Ideas

- ◆ **Have your child practice forming numbers/reading numbers on an abacus.** Because abacuses clearly demonstrate place value, they make excellent learning tools. An abacus will also come in handy when you teach addition, subtraction, and decimals. Appendix C has information on making or purchasing an abacus, as well as instructions on using one.

- ◆ **Have your child make a simple quipu (the Inca's counting device).** Although the full quipu system was extremely complicated and only special quipu makers, called quipucamayocs, were able to interpret them, your child can make a simple quipu from household items. He will need a stick and various colors of yarn or twine. (The Incas color-coded their quipus, using colors to help designate the type data recorded. Your child can use any colors he wishes.) Have your child cut the yarn into different size strands (he will need at least a yard or so in each strand), tie the strands to the stick, and he has made a simple quipu! Note: For further information on the quipu, see the picture and description on page 197.

You could have your child record different quantities on his quipu by tying knots. The place of the knot on the string determines its value. To represent 2008, tie two knots near the top of the string and eight near the bottom. Have your child record the number of people in his family, the current year, his age, the current temperature, or other numbers he wants to remember.

◆ **Teach your child to read and write Roman numerals.** Once your child is familiar with the Hindu-Arabic decimal system, have him explore Roman numerals, a system still used on buildings, clocks, and other places today. Roman numerals follow a totally different place value system. Seeing another writing system can help your child view our current way of writing numbers as *one* way of describing quantities. Appendix B contains a basic explanation of Roman numerals, along with some historical tidbits.

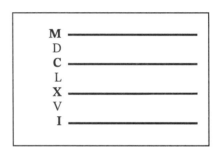

If working with an older child, you could make a counting table abacus using Roman numerals and have him practice adding and subtracting on it (see Appendix C). Counting tables were very popular in medieval times because of how easy they made working with Roman numerals. For comparison, have your child try solving math problems with Roman numerals on paper instead of with a counting device — he should quickly see that Roman numerals do not lend themselves well to written computational methods.

◆ **If your child enjoys history, have him practice writing some of the numbers in his textbook using different systems.** Appendix B offers some basic information.

◆ **Explore binary and hexadecimal numbers.** An Internet search for either "binary numbers" or "hexadecimal numbers" should turn up plenty of resources explaining these systems.

Remember: You can watch Katherine walk through teaching place value (and the other concepts) and building an abacus in the optional eCourse available at MasterBooksAcademy.com.

PARTING NOTE

As you teach place value, take advantage of those teachable moments to share God's truths. You might ask your child to think about *why* God gave us the ability to explore and communicate. Why are we, unlike the animals, made in God's image?

We were designed to know God! God has always wanted a relationship with us. He gave us the ability to communicate so we could communicate with Him! Even after we rebelled against Him, God sent His Son, Jesus, to die in our place so we could again approach Him. Through Christ, we can have access to the King of the universe.

For through him we both have access by one Spirit unto the Father.

EPHESIANS 2:18 (KJV)

Comparing Numbers

We compare all the time. We compare prices and sizes at the store, costs and revenue, and different travel options, to name a few. Even a young child is quick to point out when a sibling got "more" of something than he did.

The ability to compare is foundational to math. Why would we even bother to add, subtract, multiply, or divide if we had no ability to recognize one quantity as being different than another? If we did not know how quantities compared with each other, how could we compare the prices of items in the store ... or the cost of making a product with the revenue from selling it? How would we know if 17 or 20 represented the greater quantity?

Although we tend to take the ability to think and compare for granted, this ability is a special gift from God! As we have discussed in previous chapters, God created us. Because He fashioned us to know and communicate with Him, we can count, record, and compare quantities.

The scriptures listed below give us an even fuller picture of math's dependency upon God.

> *For in him we live, and move, and have our being.*
>
> ACTS 17:28A (KJV)

> *For who maketh thee to differ from another? and what hast thou that thou didst not receive? now if thou didst receive it, why dost thou glory, as if thou hadst not received it?*
>
> 1 CORINTHIANS 4:7 (KJV)

> *For the LORD giveth wisdom: out of his mouth cometh knowledge and understanding.*
>
> PROVERBS 2:6 (KJV)

All of us — the Christian and non-Christian alike — are dependent upon God for our very breath and abilities. In Him we "live, and move, and have our being." Even our very thought process and ability to compare numbers comes from God!

Biblical principles provide us with a framework for understanding why we can compare numbers. If our minds were generated by chance, we would have no reason to suppose we could make logical comparisons. But the Bible makes it clear — we were not designed by chance, but by a loving Creator.

Algebra ... in Second Grade?

Three symbols quite useful in comparing quantities are the greater than sign (>), the less than sign (<), and the equals sign (=). These signs are agreed upon "shortcuts" for describing how quantities compare — shortcuts children learn at an early age.

Would it surprise you to learn that these symbols are actually algebraic symbols? Although we typically associate the word *algebra* with letters, any time we use a non-numerical symbol in math, we are using algebra.

Mathematician David Eugene Smith explains:

> *The symbols of elementary arithmetic are almost wholly algebraic, most of them being transferred to the numerical field only in the 19th century,[15] partly to aid the printer in setting up a page and partly because of the educational fashion then dominant of demanding a written analysis for every problem.[16]*

There you have it — every time you draw an equals sign, you are using algebra! Since symbols fill modern textbooks, it is easy to equate math with symbols rather than viewing these symbols as useful algebraic "shortcuts." While symbols help simplify math, math can be done without them! Symbols are *not* all there is to math.

Understanding this is important! If we equate symbols with math, we could end up looking at math as some sort of independent, man-made structure rather than as a tool to describe God's creation.

Different Shortcuts

Would it surprise you to learn equality has not always been expressed using two parallel lines (=)? Figure 2 shows a few out of the many different symbols historically used to represent equality. Instead of symbols, many cultures also used words or contractions to describe equality (*pha, aequantur, aequales, gleich*, etc.[17]).

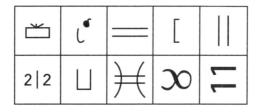

Figure 2: Some Different Equality Signs[18]

---

15. "There are, of course, exceptions. The Greenwood arithmetic (1729), for example, used the algebraic symbols." Smith, *History of Mathematics*, 2:395.
16. Smith, *History of Mathematics*, 2:395.
17. Cajori, *History of Mathematical Notations*, 1:297.
18. Top Row: One of several possible symbols the Egyptians may have used as a form of an equals sign (this particular hieroglyphic means "together" and could have been used to symbolize the results of addition); symbol used by Diophantus (200s); form of modern symbol presented by Recorde (1557); symbol used by Buteo (1559); symbol used by Holzman, better known as Xylander, that several other mathematicians adopted.
Second Row: Symbol used by Hérigone (1634); another symbol used by Hérigone; one of the symbols used by Leonard and Thomas Digges (1590); symbol popularized by Descartes (1637) — it was this symbol that proved our modern symbol's main competition; a way our modern sign was sometimes printed.
Symbols based on Cajori, *History of Mathematical Notations*, 1:297-309, and on Smith, *History of Mathematics*, 2:410-411.

Looking at different ways of signifying equality helps us realize the symbols ">,<, and =" did not just "appear" in textbooks. They are *one* set of shortcuts developed to simplify tasks! Much as abbreviations often save us time when taking notes, these mathematical abbreviations save us time when comparing quantities.

Why Bother with the Symbols?

You may wonder why we need shortcuts for greater than, less than, and equals. Is it not pretty obvious if a number is greater than or equal to another? Yes, it is. But these symbols save time and make for easier reading.

Because God has chosen to govern this universe consistently, we can record general consistencies about the way things operate. Letters and symbols help make this process easier. For example, the relationship between the power and voltage/current flowing through an outlet can be expressed as $P = V \times I$ (an abbreviation for "power equals voltage times current").

When working with equations like this and others much more complicated (God created a complex universe, and sometimes describing the consistencies He sustains requires complex equations!), using symbols rather than words saves time and makes the equation easier to read.

God's faithfulness in holding all things together consistently makes it possible for us to set up equations describing how things operate. It would be pointless to say the power equalled the voltage times the current unless electricity consistently flowed through outlets in a predictable way. Our very need for a symbol of equality testifies to God's faithfulness!

Conclusion

Comparing quantities is a gift from the Lord, the Source of all knowledge. And symbols such as $>$, $<$, and $=$ give us a shorthand way of comparing quantities on paper.

TEACHING SUGGESTIONS AND IDEAS

Objective: *To teach your child to compare quantities and view symbols as useful shortcuts, understanding the ability to compare comes from God and should be used for His glory.*

Specific Points to Communicate:
- *Comparing quantities is a gift from the Lord.*
- *Comparison symbols are shortcuts for comparing quantities on paper.*

Throughout arithmetic and even beyond, students learn to compare numbers. A young child might be shown pictures of different quantities and asked which is greater. A slightly older child might be taught to compare numbers using greater than, less than, does not equal, and equal signs ($>$, $<$, $\neq$, $=$). It is vital that the child understand the symbols $>$, $<$, $\neq$, and $=$ are symbols we have adopted to help us communicate about a number's size (note that on the $>$ and $<$ than signs, the larger end should be facing the larger number).

Example

Below is a very typical presentation of the less than and greater than signs.

> *Count the rabbits and the frogs.* [There are pictures of a group of 7 and a group of 5 rabbits, and a group of 10 and a group of 6 frogs.] *Write the numbers. Complete the number sentence. < means less than. > means greater than.*[19]

Like most textbook presentations, this presentation states several truths. The symbol < does mean less than, and > does mean greater than. But the presentation does not make it completely clear these symbols are *one* shorthand way of comparing quantities on paper. Students reading it could easily get the wrong view of math if they learn to view math as an *independent* set of symbols and facts rather than as one method of describing God's creation.

The teacher's key lesson designed to go with the presentation above attempts to bring God into the lesson by discussing coyotes and trusting God with our fear.

> *Discuss how the students would feel if they heard coyotes while they were camping one night. Remind the class that even though they might feel a little afraid, they should always trust in God...Display "Coyote Sign >" and "Coyote Sign <".... Call attention to the mouth of each coyote and point out that it is shaped like the less than sign and the greater than sign....*[20]

While we should trust God at all times and not be afraid, discussing this does not really do anything toward building a biblical worldview of the less than and greater than signs themselves. Students are being taught to trust God when afraid, but still missing out on realizing God's presence in math.

Let us take a look at one way we could present our current greater than and less than signs as useful ways of comparing real-life quantities. Notice how the following presentation leaves the child looking at these symbols as useful shortcuts. The presentation encourages the child to view math as a God-given tool (something *dependent* on God) rather than as an *independent* fact.

> *Today we are going to learn a little about comparing quantities. Count the rabbits here. Obviously, this group has more rabbits in it. God gave us the ability to recognize that seven represents a larger quantity than five.* [Demonstrate with manipulatives and other examples if necessary.]
>
> *Can you think of how to express this on paper? You already have learned symbols to represent seven and five (7 and 5). But how could you easily show that seven represents a larger quantity than five?* [Listen and respond as appropriate.]
>
> *We could write "7 is greater than 5." When we write something over and over again, though, it is wise to simplify as much as possible. Because comparing quantities helps us learn about and express different aspects of God's creation, we find it helpful to use*

---

19. Jacobs, et al., *Math 2 for Christian Schools: Teacher's Edition*, 53 (in a quote from the student version).
20. Jacobs, et al., *Math 2 for Christian Schools: Teacher's Edition*, 52.

symbols for "greater than" and "less than." These symbols make it easier to express how numbers compare, saving us from writing extra words.

While people have used different symbols, today we use two basic symbols called the greater than sign (>) and the less than sign (<).

Now, there are lots of other ways you could present the greater than and less than signs. You could even teach them without a textbook! However you choose to present this concept, remember to help your child see these symbols as useful shortcuts we can use because of the ability God gave us.

Ideas

- ◆ **Have your child measure or count items around the house and see which distance/quantity is greater or less than the other.** Real-life applications go a long way in taking math out of a textbook and showing how it describes the quantities around us.

- ◆ **Draw a number line and have your child use it to compare numbers, or have him compare dates on a timeline.** Number lines serve as visual aids, helping us see which numbers are greater than or less than others. Timelines also help compare numbers (historic dates and periods); you may want to pull out a history book and take a look at a few timelines with your child. You could even look at a timeline of the kings of Israel and point out how comparing dates helps us see when different biblical events occurred.

- ◆ **Have your child compare a variety of quantities on paper without using symbols.** Help your child realize the sense in abbreviating words or phrases we repeat over and over by having him write comparisons in words. You could also show your child other symbols men have used, or have him try to think of his own symbols. These sort of activities help present a biblical understanding of math by leaving the child with a clear understanding that <, >, and = are one set of symbols rather than a fact to rotely memorize.

PARTING NOTE

Since God is the Source of *all* knowledge, we can trust Him and seek His wisdom in *every* area of life. Nothing is outside His jurisdiction.

Addition:
Foundational Concept

$$2 + 2 = 4 \qquad 3 + 2 = 5 \qquad 4 + 2 = 6$$

Addition facts appear neutral enough, don't they? Yet these little "facts" loudly proclaim God's praises!

All around us, we observe these "facts" in action. If we take two objects and put them together with two other objects, we consistently end up with four objects. Why is this? Why can we rely on the facts we memorize in a textbook to actually end up working in real life? Why can we memorize what 2 + 2 equals and expect it to apply in a wide variety of situations?

Math's ability to work has puzzled mathematicians and philosophers for centuries. It does not make sense from a human standpoint why math works outside a textbook. Unable to find an explanation, most scientists and math books ignore the question of why math works.

> *The vast majority of working scientists, myself included, find comfort in the words of the French mathematician Henri Lebesgue: "In my opinion a mathematician, in so far as he is a mathematician, need not preoccupy himself with philosophy."*
>
> FREEMAN J. DYSON[21]

However, we do not need to ignore the vital question of why math works. The Bible offers an explanation! Let us take a look at a few biblical principles together, examining especially how they apply to addition.

Biblical Principles

The Bible teaches us God both created all things and holds them together (Colossians 1:16, 17). He is a consistent, never-changing God (Malachi 3:6), and He has chosen to hold this universe together in a consistent way — according to "fixed laws."

> *"This is what the LORD says: 'If I have not established my covenant with day and night and the fixed laws of heaven and earth, then I will reject the descendants of Jacob and David....'"*
>
> JEREMIAH 33:25-26A (NIV)

We can rely on addition facts to work because they describe on paper the consistent way God causes objects to interact! *Because* God has chosen for objects to combine predictably, we can say with confidence, "Two plus two will equal four." Addition describes a "fixed law" God has put in place.

---

21. Dyson, "Mathematics in the Physical Sciences," *Mathematical Sciences*, 101–102.

Just think. God holds this universe together so consistently we can memorize how objects will add! Addition facts like 2 + 2 = 4 are only useful because, day in and day out, God keeps His covenant with the "fixed laws of heaven and earth" — laws like addition, subtraction, multiplication, and division, as well as gravity and scientific "laws." Addition facts are really a record of God's faithfulness and consistency.

Every time we add, addition can remind us that, just as God is faithfully keeping His covenant with the "fixed law" of addition, He will be faithful to everything else He has promised. He is a faithful, trustworthy God.

Making Sense of the Apparent Inconsistencies

You may be thinking, *Wait a minute! What about miracles? Jesus fed 5,000 with 5 loaves of bread and 2 fish. Those 5 loaves of bread certainly did not add in a typical manner! How can we say God faithfully holds things together consistently if He performs miracles?*

God is consistent *to His nature*. In His mercy and faithfulness, He governs this universe in a predictable pattern. Out of His same mercy and faithfulness, He sometimes chooses to perform a miracle. Miracles help us see God's sovereignty. Miracles show us God is the One in charge!

Addition has another apparent inconsistency. To see it, simply dip your hand in a cup of water and let a droplet fall onto a plate. Dip your hand again and drop another droplet of water on top of the first one. What happened?

The droplets should have merged together into a single droplet of water. Although one water molecule plus another equals two, one droplet of water plus one droplet of water does *not* equal two droplets — they merge together to form one larger droplet. This seems like a huge inconsistency — why doesn't 1 water drop + 1 water drop = 2 water drops?

This apparent inconsistency makes sense when we think about it from a biblical worldview. God could have different, though equally consistent, principles for governing liquids than He does solids. The apparent contradiction in how liquids combine reminds us that, no matter how well we think we have things figured out, God's laws and universe are more complex than we can imagine!

On the other hand, the way raindrops combine presents a huge problem if we believe math is a product of human reasoning. After all, human reasoning says 1 + 1 = 2. Obviously, human reasoning cannot be trusted when dealing with raindrops. And if not here, can we trust any of math? Or is it all just a delusion?[22]

Whenever we abandon a biblical worldview, we end up with a host of difficulties. Thankfully, though, God has given us His truth to live upon! If we take Him at His Word, we can use addition confidently, knowing it does not rest on man's reasoning, but on God's

---

22. On page 92 of *Mathematics: The Loss of Certainty*, Morris Kline shares how the raindrop quandary, along with many others, was brought up by Hermann von Helmholtz in 1887 and explores the problem it posed to those who had enthroned human reason and math. While *Mathematics: The Loss of Certainty* does not approach the topic from a biblical perspective, it nonetheless clearly shows the complete loss of certainty man is forced to adopt when he abandons a biblical basis for mathematics.

power. At the same time, we can recognize God's greatness and expect to see Him holding this universe together in a marvelous way no one set of math facts could ever adequately describe.

Different Ways of Recording Addition

As we discussed in the last chapter, when we have to write something over and over again, it helps to develop symbols or "shortcuts." Most of us use the symbol "+" as a shortcut for addition. But this symbol, often referred to as a plus sign, is not the only symbol for addition! A look at history shows us different cultures have recorded addition different ways. The principle God created has remained unchanged, but over the years men have used their God-given creativity to express it differently.

Many cultures throughout history did not use a sign at all. They used other methods to signify addition, such as a number's location in relationship to another number, or words (example: Add 2 to 4). Others used very different signs from what we use today. The Egyptian Ahmes papyrus used a picture of legs walking forward,[23] the Greeks sometimes used a line,[24] and some mathematicians used a p or a variation of p.[25] The Bakhshali *Arithmetic* uses a *yu* to represent addition and actually uses the + sign to signify a quantity you would subtract.[26]

Our current plus sign (+) seems to have begun in Germany sometime between 1450-1500 A.D. It probably came from the word *et,* which meant *and.*[27]

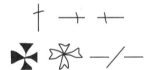

Even after our current plus sign arrived on the scene, it took a long time for everyone to adopt it. Along the way, people used different variations. A few other variations are shown.[28]

God's principles do not change, although the way of recording them can (and does!). Addition is addition, no matter what symbol we use to describe it. Symbols only *represent the reality of addition God created and sustains.*

---

23. Ahmes papyrus, problem 28. Quoted in Cajori, *History of Mathematical Notations,* 1:230, as from T.E. Peet, *The Rhind Mathematical Papyrus,* Plate J, No. 28. As is often the case with math notation, the use of this symbol was not universal; Cajori notes on page 229 that a different Egyptian papyrus uses a pair of legs to represent squaring a number instead of adding. David Eugen Smith notes that the actual Ahmes papyrus was written right to left, meaning the symbol would have been the reverse of what is pictured, appearing as a pair of legs walking backwards when read from left to right. See Smith, *History of Mathematics,* 2:396.
24. Cajori, *History of Mathematical Notations,* 1:229-230. Cajori also notes that Diophantus, a Greek writer from the third century A.D., used juxtaposition (placing numbers next to each other) instead of a line.
25. For example, p´ was one of the symbols English mathematician Oughtred used to represent addition. See Cajori, *History of Mathematical Notations,* 1:190, in a table of Oughtred's Mathematical Symbols "first published, with notes, in the *University of California Publications in Mathematics*, Vol. 1, No. 8 (1920), p. 171–86." Italian mathematicians in the 1500s also used this symbol (p. 230).
26. Cajori, *History of Mathematical Notations,*1:230. See also Midonick, ed., *Treasury of Mathematics,* 93, and Smith, *History of Mathematics,* 2:396.
27. Cajori, *History of Mathematical Notations,* 1:230, and Smith, *History of Mathematics,* 2:398.
28. Cajori, *History of Mathematical Notations,* 1:236–239.

Conclusion

When you teach 2 + 2 = 4 and other addition facts, let your heart pause and ponder the miracle of addition. Our God is governing this universe together so consistently we can memorize how objects will interact. Even on those days when things are not going according to our plans, we can still take two objects, add two more, and end up with four because God is *still* in charge. He has not been — and never will be — knocked off His throne.

Is it not mind-boggling to realize the very addition facts we memorize as young children only have real-life meaning because of God's faithfulness?

TEACHING SUGGESTIONS AND IDEAS

Objective: *To help your child see that God causes objects to operate orderly and consistently enough for us to memorize how quantities combine, and to guide him into learning his addition facts while at the same time applying them in a God-honoring fashion.*

Specific Points to Communicate:

- *God holds all things together so consistently we can memorize how quantities will add.*
- *Symbols like the plus sign (+) are shortcuts for representing addition.*

Addition is usually the very first math operation students learn. As such, addition lays the groundwork for how a child learns to view math operations in general. So be careful about getting lost in mere fact memorization! While learning addition facts is very important, it is even more important for your child to understand those facts describe consistencies God holds in place and therefore serve as a great aid in completing the tasks God has given us to do.

Connecting addition facts with real-life settings plays an important part in laying this foundation. Try showing your child the consistent way objects add in a real-life setting both before making him memorize his addition facts and throughout the memorization process. Keep demonstrating how addition facts describe the consistent way God makes objects combine. While you will also want to help your child learn addition facts well enough to recall them quickly, remember to show the "big picture."

Example

Below is a typical addition presentation.

> *"Make a tower of two and another tower of two."*
>
> ..."*Put your towers together."*
>
> *"How many cubes do you have now?"*
>
> *"Two and two is four."*
>
> Write "2 + 2 = 4."[29]

---

29. Larson, with Mathews and Wescoatt, *Math 1*, Home Study Teacher's ed., 126–127.

Notice the presentation shows students addition facts work by using tower manipulatives, but does not ever explain *why* addition works or praise God for His faithfulness in holding all things together consistently.

We can modify this presentation so as to convey these missing elements in a similar way we would modify a science lesson that looked at a leaf and explained the intricacies of photosynthesis without ever giving glory to the Creator. Much as we would pause and realize God created the leaf's complex structure, praising Him for His handiwork, we can pause and realize these addition facts describe consistencies God sustains, and praise Him for His handiwork.

One way to accomplish this would be to start by having your child add various objects, such as pencils, raisins, or toys (or have him add the towers or manipulatives your math book suggests). Point out that if he starts with two and combines it with two more, he consistently ends up with four, and ask him why this is. Then proceed to introduce addition.

> *Two plus two consistently equals four because God chose to make it that way! God decided how objects would combine together, and, day after day, year after year, He keeps them combining together that way.*

> *There is a fancy word to describe combining things together. That word is addition. When we combine objects together, we say we are adding them.*

You could then introduce how we represent addition on paper, explaining that the plus sign (+) is just one way to express addition.

When it comes time to begin memorizing addition facts, you might say something like this:

> *God is holding this universe together so consistently we can memorize that two of something added to two more will equal four. If we memorize this, next time we need to add two and two we will already know the answer. Not only can we see what two and two will equal, but we can memorize what two and three or two and four will equal too! We can memorize how objects add because God is faithful.*

You could read Hebrews 1:3 together and talk about how Jesus is "upholding all things by the word of his power" — night and day, He is controlling how objects add. And He does it so faithfully we can memorize addition facts!

Do you see how these little changes give your child a completely different view of addition? Rather than memorizing facts without understanding why those facts work (which could leave him thinking mathematicians are sure smart), he can now memorize facts while realizing that these facts very existence testifies to God's faithfulness.

There are many, many other ways you could present addition, but hopefully those ideas and the ones listed in the next section will get you started. As you approach addition as something dependent on God and His faithfulness in holding all things together, you will be amazed at how many opportunities present themselves to foster a spirit of worship while learning math.

Ideas

- ◆ **Use household objects and situations.** Having your child apply addition to real-life situations can help him view addition facts as ways of describing real-life consistencies rather than as empty facts confined to a textbook. While we will look at more addition applications when we explore multi-digit addition, here are a few to get you started.

  - » **With Younger Children** — If you have a younger child, why not take advantage of his play to teach him to view math as a useful tool? For example, say your child likes to play with stuffed animals. One day you notice he has four stuffed dogs lined up along the wall in preparation for taking them for a "walk." You could turn this into a math lesson simply by asking him how many animals he is taking for a walk (four). Then hand him another stuffed animal, asking him how many he has now (five). Explain that, because God is so consistent, four plus one consistently equals five, making it a very useful fact to memorize.

  - » **With Older Children** — As you find yourself using addition, mention it to your child or have him apply whatever addition facts he has learned. Because God holds all things together (not just those apparently more important) and because addition is a way of recording His creation, addition can be used in thousands of everyday circumstances. For example, when we have company, we often mentally add up the people in each family to figure out how many chairs and place settings we need. If the Smith family has 6 members, the Patterson family 5 members, and our own family 7 members, then we know we need 6 + 7 + 5, or 18 chairs and place settings. Keep your eyes open for other examples.

- ◆ **Have your child practice adding on an abacus.** Using an abacus is a wonderful way to visually "see" addition. Abacuses also reinforce place value — and can provide hours of fun! Appendix C has instructions on building (or purchasing) and using an abacus.

PARTING NOTE

Just as God is faithful to hold this universe together, He will be faithful to *everything* else He says in His Word. He will be faithful to judge all who have rejected His Son (2 Thessalonians 1:6, 8,9), and faithful to save those who have come to Him through Christ Jesus (Acts 2:21; 2 Thessalonians 1:6, 8–9). We serve a God we can believe and rely on completely.

Subtraction:
Foundational Concept

$$12 - 5 = 7 \qquad 3 - 2 = 1 \qquad 10 - 5 = 5$$

If you start with $12 and spend $5, you will end up with $7. If you start with a 12 inch board and cut 5 inches away, you will have 7 inches left. If you start with 12 lessons and finish 5, you will have 7 left undone. Twelve of something minus five consistently equals seven, making it possible for us to call this consistency a fact and memorize it.

We take subtraction for granted and use it all the time in numerous situations. But why does subtraction work? Because, like addition, it is a *way of expressing a consistency God created and upholds!* The Bible makes it clear Jesus Christ, God the Son, created and upholds all things (Colossians 1:16–17; Hebrews 1:3). Subtraction facts record the consistency He sustains. If it were not for the consistent, faithful way God has chosen to hold this universe together, subtraction facts would be absolutely meaningless outside a textbook.

Is it not ironic that, although subtraction would be meaningless without God, most textbooks do not even mention Him or give Him credit for subtraction? Why is God so often ignored?

Once again, we find the Bible gives us the answer.

What Happened?

The first chapter of Romans tells us all of creation clearly declares God's eternal power and Godhead. While we usually think of creation here in terms of science, this passage also applies to math, as math records the consistencies throughout creation — consistencies testifying to God's power in holding things together.

> *For the wrath of God is revealed from heaven against all ungodliness and unrighteousness of men, who hold the truth in unrighteousness; Because that which may be known of God is manifest in them; for God hath shewed it unto them. For the invisible things of him from the creation of the world are clearly seen, being understood by the things that are made, even his eternal power and Godhead; so that they are without excuse:*
>
> ROMANS 1:18–20 (KJV)

Although men can clearly see God's power, they choose to ignore it and give His glory elsewhere.

> *Because that, when they knew God, they glorified him not as God, neither were thankful; but became vain in their imaginations, and their foolish heart was darkened. Professing themselves to be wise, they became fools, And changed*

the glory of the uncorruptible God into an image made like to corruptible man, and to birds, and fourfooted beasts, and creeping things.

ROMANS 1:21–23 (KJV)

It is not hard to see how men have not glorified God in science. Many have adopted the theory of evolution and have sought to give chance and the "laws of nature" credit for our very existence, thereby giving God's glory to matter and natural processes.

A similar ignoring of God occurs in math. Rather than acknowledging God as the Creator and Sustainer of the mathematical consistencies around us and being thankful, many have ignored Him and viewed math as an independent fact, thereby giving glory for the consistencies around us to "mathematical laws" of nature or to man's own abilities to find order amidst a presumed chaotic universe. In math, God's glory often gets attributed to nature, math itself, or man.

Below are a few quotes illustrating this point. Notice how each quote attributes the credit for math to something or someone other than God.

> *Mathematicians are architects of complex systems.*[30]
>
> *Many physicists have been so impressed by the usefulness of mathematics that they have attributed to it almost mystical power.*[31]
>
> *One cannot escape the feeling that these mathematical formulae have an independent existence and an intelligence of their own, that they are wiser than we are, wiser even than their discoverers, that we get more out of them than was originally put into them.*[32]
>
> *Natures' laws are man's creation. We, not God, are the lawgivers of the universe.*[33]

It is not that math does not clearly declare God's praises — it does! But men have chosen to be "willingly ignorant" (2 Peter 3:5). Even Einstein, an atheist himself, confessed to something miraculous in math atheists could not explain.

> *Even if the axioms of the theory are posited by man, the success of such a procedure supposes in the objective world a high degree of order which we are in no way entitled to expect a priori. Therein lies the "miracle" which becomes more and more evident as our knowledge develops.... And here is the weak point of positivists and of professional atheists, who feel happy because they think that they have not only pre-empted the world of the divine, but also of the miraculous.*[34]

---

30. Guedj, *Numbers: The Universal Language*, 74.
31. Dyson, "Mathematics in the Physical Sciences," *Mathematical Sciences*, 97.
32. Heinrich Hertz (German physicist; on complex numbers in quantum mechanics) quoted in Dyson, "Mathematics in the Physical Sciences," *Mathematical Sciences*, 99.
33. Kline, *Mathematics: The Loss of Certainty*, 98.
34. Albert Einstein, *Letters À Maurice Solovine* (Paris: Gauthier-Villars, 1956), 114–115, quoted in Nickel, *Mathematics: Is God Silent?* 210, as a quote from Stanley L. Jaki, *The Road of Science and the Ways to God* (Edinburgh: Scottish Academic Press, 1978), 192–193.

Day after day, we see God's hand all around us. We see the consistency and order He is keeping together throughout creation. How sad when we ignore Him and look at math as independent facts — as a complex system man engineered rather than as a description of consistencies God created and faithfully sustains!

Despite how we ignore God, He lovingly continues to draw men to Himself.

> *The Lord is not slack concerning his promise, as some men count slackness; but is longsuffering to us-ward, not willing that any should perish, but that all should come to repentance.*
>
> 2 PETER 3:9 (KJV)

Even though we constantly forget God, He never forgets us. Day after day, He faithfully keeps all things together. Day after day, He urges us to come to Him — to admit our inability and sin and receive the righteousness He offers us freely in Jesus Christ.

Conclusion

Think about the miracle of subtraction for a moment. Every day of our lives, we rely on subtraction to work. We know money will not mysteriously reproduce itself in our wallets. Yet subtraction is only consistent because the God of the Bible — a consistent, never changing God — upholds all things.

In an effort to ignore God, men often ascribe His glory to themselves or nature. Yet God remains faithful, even when we ignore Him. His testimony remains for all to see within the very subtraction facts we memorize.

TEACHING SUGGESTIONS AND IDEAS

Objective: *To continue to help your child see God causes objects to interact orderly enough for us to memorize how quantities combine. To guide your child into both memorizing and applying his subtraction facts.*

Specific Points to Communicate:
- *Subtraction expresses a consistency God created and upholds.*
- *Symbols like the minus sign (-) are agreed-upon shortcuts for representing subtraction.*

Flashcards, time tests, subtraction games — there are many different ways to go about helping your child memorize his subtraction facts. Yet your goal goes beyond mere fact memorization. You want your child to see each one of those facts as a testimony to the fact that God is in charge. You want him to be able to use those facts to serve the Lord.

How do you reach those goals? Start by connecting subtraction facts with real-life objects. Show your child we can only able to memorize all these "facts" because God is a faithful, all-powerful God. Look for simple opportunities to demonstrate the principle of subtraction using household items, thereby continuing to present subtraction as a useful tool.

Example

As with addition, typical subtraction presentations teach students that subtraction works but fail to explain *why* it works. Below is an example of this. In the actual curriculum, colorful graphics and illustrations accompany this very factual text.

> *There are 8 birds. 3 fly away. 5 birds are left.*
>
> *We write the number sentence: 8 - 3 = 5*
>
> *This is **subtraction**. It means **taking away**.*
>
> ***Subtract** 3 from 8. The answer is 5.*[35]

Although the curriculum goes on to give a lot of different subtraction pictures and examples, it does not teach the student *why* subtraction works. The student is left looking at subtraction as a fact *independent* from God rather than *dependent* upon Him.

There are lots of different ways you could rework subtraction presentations to give your child a different view of subtraction. Rather than saying, "This is subtraction," you might say something like, "Subtraction is the fancy name we use for taking or giving away objects." Rather than saying, "We write the number sentence," you might say something like, "Let's learn one way we could represent subtraction on paper." And rather than saying, "The answer is 5," you might say something like, "Because of the consistent way God holds things together, we can rely on ending up with 5 if we subtract 3 from 8."

Do you see how even these little changes give a different perspective on subtraction? Of course, there is a lot more you could do! You could talk about how we can only memorize how objects will subtract because they subtract the same way all the time. You could then take your child to the Bible and discuss a characteristic of God subtraction helps us see, such as how He is all-present. Below is an example.

> *Although the word "subtraction" is not found in the Bible, the Bible gives us the only framework for understanding why subtraction works that makes sense. It tells us in Jesus Christ, "all things hold together" (Colossians 1:17 NIV). Subtraction works because Jesus holds everything — from the tiniest atom to the largest galaxy in space — together consistently.*[36]
>
> *I find this thought so very encouraging. No matter where I am, I can take 4 objects, subtract 1, and be left with 3. Why? Because no matter where I am, Jesus is there too, holding all things together consistently and causing objects to subtract in the same, predictable way.*

---

35. Primary Mathematics Project Team, *Primary Mathematics 1A*, Textbook, 39.
36. This does not mean God is bound by subtraction — He can do anything He wants! And who knows? We may discover He has chosen to hold aspects of creation together in a different, though equally consistent, fashion. The point remains: God is in control of everything. Because of His sustaining hand, we are able to memorize subtraction facts and use them with confidence in thousands of situations and places.

The Bible makes it clear Jesus is one with God the Father (John 1:1; John 14:9). It also shows us over and over again that God is in charge of everything everywhere — we cannot run away from God.

> *Whither shall I go from thy spirit? or whither shall I flee from thy presence? If I ascend up into heaven, thou art there: if I make my bed in hell, behold, thou art there. If I take the wings of the morning, and dwell in the uttermost parts of the sea; Even there shall thy hand lead me, and thy right hand shall hold me.*
>
> PSALM 139:7-10 (KJV)

Subtraction reminds us of this truth. Nothing in our lives or our hearts escapes God's notice. Nothing happens to us without His okay. No circumstance, no matter how bad it might be, can keep us from God! We are never out of God's care. No matter where we are, He is there. If you ever have any doubt of His presence, grab 4 objects, take 1 away, and look at the 3 remaining. Let subtraction encourage you that no matter what your circumstances, God is still in charge, faithfully holding all things together.

Note: You could have easily chosen a different attribute of God, such as His faithfulness in holding this universe together or the power of His Word (a Word powerful enough to uphold everything consistently).

When it comes time to present terms like *minuend*, *subtrahend*, and *difference*, try rewording the presentations to present the terms as names rather than as facts. A minuend is a name we use to help us refer to the starting quantity of items we have. If we start with $40 and spend $20, we would call $40 the minuend. Names like *minuend* allow us to quickly describe quantities. "Minuend" is a lot simpler to say than "the quantity I started with." Presenting terms as names teaches the child to view these terms as useful tools rather than as truths in themselves.

In short, approach each aspect of subtraction as dependent on God, and let our very ability to memorize how objects will subtract encourage both you and your child that you *can* trust God.

Ideas

- ◆ **Connect subtraction facts with real-life objects.** Connecting facts with objects will help your child see subtraction as a way of describing a real-life consistency God created and sustains rather than as an independent fact. Try demonstrating an equation using an abacus, manipulatives, or any household objects. (If you have not done so yet, please follow the instructions in Appendix C to build/buy and familiarize your child with an abacus. Abacuses also go a long way in making subtraction fun and easier to learn — especially for your hands-on learner!)

- ◆ **Read Romans 1 and discuss the biblical and non-biblical views on math.** You could also look at Genesis 3 and discuss how ever since the Garden of Eden, satan has been trying to get man to trust his own reason instead of God's Word. But God's Word is never wrong — it, and it alone, gives us a logical explanation for math's very existence.

- **Pick a mathematician to study.** Looking at a mathematician's beliefs can be a great way to talk about beliefs toward math. See Appendix A for some ideas to help guide your study.

PARTING NOTE

Everyone can see subtraction works. We subtract on paper and apply our answers to real-life settings, confident real-life objects subtract consistently. We keep check registers with assurance, knowing if we add and subtract correctly, the balance in our checkbook will be the same as if we could have physically counted the money. Everything around us clearly operates according to consistent principles.

But not everyone sees subtraction the same way. We either look at subtraction as dependent upon God, or as independent from Him. We either give the glory for subtraction's consistency to God, or else ascribe His glory to something else, such as mathematical laws or our own human reason.

As you study subtraction, let it encourage you to trust the God of the Bible.

Addition & Subtraction:
Multi-Digit Operations

We cannot possibly memorize every number we will have to add or subtract in our lives. We need a method for adding and subtracting numbers we do not have memorized.

In the method typically taught today, we put the numbers we want to add or subtract on top of each other and add or subtract each column individually, working from right to left and carrying or borrowing to/from the next or previous column as necessary. If we do everything correctly, we will end up with an accurate answer.

$$\begin{array}{r} \overset{1}{1}2 \\ +19 \\ \hline 31 \end{array} \qquad \begin{array}{r} \overset{11}{\cancel{2}1} \\ -14 \\ \hline 7 \end{array}$$

Pretty cool — but what makes this method work? What is really happening here? Let us delve past the mechanics and "reveal" multi-digit addition and subtraction methods.

Different Methods to Express the Same Consistency

History helps in "revealing" addition and subtraction methods. Seeing other ways to add and subtract can prompt us to begin to look beyond the typical way of adding and subtracting and view these methods as ways of describing a consistency God created and sustains.

All throughout history, men have used different methods to add and subtract. Many of these methods did not involve memorizing facts or writing at all!

People have used their fingers to add and subtract — and even to multiply! By employing place value, fingers can be used to work with both small and large quantities, including numbers like 10,000 or even 100,000.[37]

At one time, the Chinese used *Sangi* boards, a flat board with both horizontal and vertical lines forming equal-sized boxes, combined with tiny sticks. They would use the board to keep track of place value, and the sticks to form numbers on the board.

In the Middle Ages, Europeans predominately used abacuses or finger reckoning to solve problems. After all, Roman numerals do not lend themselves well to written methods. In an abacus, some sort of marker (bead, pebble, or other small object) is moved along either a rod, wire, line, or groove. Someone proficient with the use of abacuses could easily add up totals as quickly as a person on a calculator can today.

As communications opened with the Arab world, Europeans learned of the Hindu-Arabic method of working with numbers. This method not only employed a different system of writing (the Hindus used our current digits — 0,1,2,3,4,5,6,7,8,9), but also made use of a different method of adding and subtracting numbers. The Hindus/Arabs added and subtracted numbers using written methods (often writing with a stick in the sand).

---

37. See Smith, *History of Mathematics*, 2:196–202, for an explanation of finger reckoning.

Even after adopting Hindu-Arabic numerals, not everyone solved addition and subtraction problems the way most of us do today. Mathematicians tried to develop different written methods, called algorithms, people could follow to quickly solve problems. These methods did on paper what abacuses did with rows and columns, making addition and subtraction simple by keeping track of place value.

Figure 3 shows some of the different methods used to add and subtract — some are quite different than the method typically taught today! (See Appendix D for an explanation of each of these methods.)

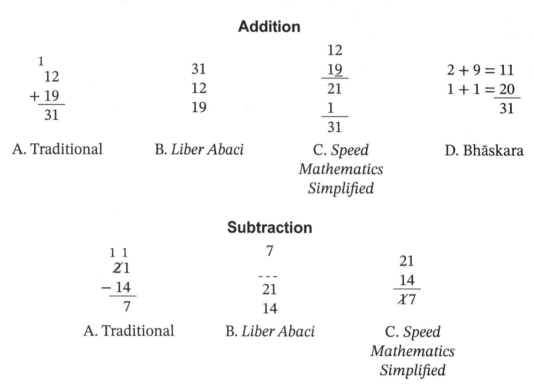

Figure 3: Different Addition and Subtraction Methods

Methods Describe Consistency

Why all these methods? What is the purpose of adding and subtracting? We learn addition and subtraction methods because they aid us in real life! They are incredibly useful tools in thousands of situations.

Yet addition and subtraction methods are only useful because they describe a real-life consistency. To better understand this, let us take a look at how a few methods describe the real-life way objects add and subtract.

Figure 4 shows adding on one form of an abacus (there are many different types of abacuses and variations within those types). Counters were "put on" or "added" to the board to represent adding 9. The word *add* actually stems from the Latin word *addere*, meaning "to put to."[38] In subtraction, counters were taken, or drawn, away. The word

---
38. *The American Heritage Dictionary of the English Language*, 1980 New College Edition, s.v. "add."

subtract comes from the Latin word subtrahere, "to draw away."[39] The counters basically served as ways of representing what would happen if quantities were physically added and subtracted.

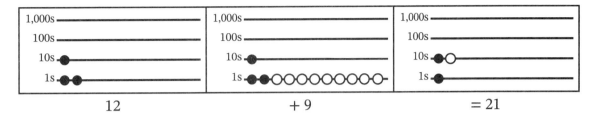

Figure 4: Adding on an Abacus
(See Appendix C for details on adding on an abacus.)

We do something very similar when we add and subtract numbers on paper. First, we add the one's column much as we would add the beads on the one's row of an abacus. We end up with 11. Much as we would replace 10 beads with 1 bead on the ten's row, we write a 1 in the ten's column. We then add up the ten's column, and get 2. Now we have our answer: 21.

Both the counting table and our modern paper method, like all methods, *assume consistency*. In order to use any addition and subtraction method outside of a textbook, we have to make these two presuppositions:

1. quantities will add and subtract in a consistent manner, and
2. man can observe and record this consistency.

As we have discussed in previous chapters, the Bible gives us a logical basis for these presuppositions. *Because* Jesus is "upholding all things by the word of his power" (Hebrews 1:3), the universe teems with consistency and order. *Because* God made man in His image, man has the ability to observe and record this consistency and order.

Adding and Subtracting with a Purpose

If we only learn to add and subtract in a textbook, we could easily end up viewing math as a mere intellectual pursuit. Math, however, is much more than an intellectual pursuit! Math methods — including addition and subtraction methods — express real-life consistencies. As such, they serve as useful tools in a variety of settings.

We employ addition and subtraction while completing many household tasks, including shopping, balancing a checkbook, and planning. No matter what our occupation, we will likely have to add and subtract to some degree or another on the job. Think of addition and subtraction as pocket knives — handy tools we utilize quite frequently!

Yet why would addition and subtraction work in real life if man developed math? Or why would we expect the universe to be consistent if it were a product of chance? Addition and subtraction's usefulness does not make sense in a secular worldview.

---

39. Ibid., s.v. "subtract."

The Bible gives us a basis for looking at addition and subtraction methods as a tool. We expect the universe to be consistent because a consistent God holds it together.

Although both Christians and non-Christians use addition and subtraction practically, only the Christian has a logical explanation for *why* addition and subtraction apply in real life.

Conclusion

Different addition and subtraction methods work because — and only because — they describe real-life consistencies God created and sustains. Because they describe real-life consistencies, we can employ them as "tools" in a variety of situations.

TEACHING SUGGESTIONS AND IDEAS

Objective: *To help your child learn addition and subtraction methods from a biblical perspective so he will be equipped to apply them in various real-life settings.*

Specific Points to Communicate:

- *The addition and subtraction method being taught is not the only way to add and subtract, but rather one useful way of expressing on paper a consistency God created and sustains.*
- *Addition and subtraction methods work because they describe real-life consistencies.*
- *Addition and subtraction methods are useful tools, not intellectual games.*

When teaching multiple-digit operations, there are a lot of rules to cover. Textbooks usually stress these rules and walk through operations in a very here-is-how-you-do-it-now-do-it fashion. As a result, students are left confused. Why do they need to carry or borrow a number? What makes these rules work? Who developed these rules anyway?

If not directly given the answer to these important questions, students learn to trust rules themselves. They see rules as the mechanisms making math work instead of as tools to help them record the way God causes things to add and subtract.

They are also not being taught math in such a way that they will be able to apply what they know to other situations. A child who knows rules, not principles, is not going to know what to do when he encounters a type of problem he has not seen before. A child being taught rules can use math in limited circumstances, but he will be much better equipped to apply the principles to unexpected tasks God may one day give him if taught the *principle* and how to think through concepts.

Instead of teaching your child to mechanically follow a rule, show him subtraction and addition in real life. Teach him how to see the "why" behind the rules and to view the rules and paper methods as *ways of describing the consistencies God sustains around us*.

Then keep reinforcing the purpose behind learning addition and subtraction methods by giving your child opportunities to apply the rules he learns to real-life settings. Part of your goal is for him to learn to use addition and subtraction to serve the Lord in whatever task he encounters — whether around the house or in a future occupation.

Learning to add and subtract multiple digits is a process — and it most likely will not happen overnight. Do not be discouraged if things move forward slowly — let your child learn at his own pace. He learned to walk at his own pace, and he will learn to add at his own pace too. God holds us accountable to be diligent and do our best, not to progress at the same pace as everyone else. Some may learn fast, and others more slowly — and that is okay! If your child is not grasping a concept, ask God for a different way of explaining it, or ask Him if you need to hold off on presenting it for a time. Listen to His leading, and do not feel tied to what "most" people do.

Example

You may find your textbook already does a good job showing how addition and subtraction methods work. Many try, at least at first, to illustrate how the method works — how it helps us keep track of place value so we can easily add multi-digit numbers on paper. Some do a better job at this than others.

For example, one curriculum presents addition using both graph paper squares and an expanded written notation that emphasizes place value. After laying this foundation, it then introduces the notation we typically use, calling it "compact notation."

> *Today I want to show you how to find the sum and difference using compact notation. I want you to write the problem in expanded notation....*
>
> *Now let us focus on the same problem in compact form.*[40]

This method leaves the child with a knowledge of the possibility of different notations and an understanding of place value, both important points to make sure your child has grasped.

Other curriculums focus only on our typical notation, emphasizing the mechanics of addition. If your curriculum presents addition and subtraction methods in more of a here-is-the-rule-to-follow-now-learn-it method, try to help your child understand why the rule works. You might have him add and subtract on an abacus, then show him how the paper method lets him do on paper the same thing he did on the abacus. Or you might use whatever manipulatives your curriculum suggests to demonstrate a little more how the paper method describes a real-life principle.

You do not want your child to get the idea the methods in his textbook *are* addition and subtraction. You do not want him to begin viewing math as an intellectual exercise. It is all-too-easy to get into this mindset if a child only uses one method and only solves problems in a textbook.

The section below offers a variety of simple ideas. As always, you could do more, so be creative and have fun.

---

40. Quine, *Making Math Meaningful Level 2, Parent/Teacher Guide*, 86 (Activity 4D).

Ideas

♦ **Look at different ways to add and subtract.** Make an abacus, or have your child try an online abacuses (see Appendix C). If you have an older child, you could even have him try using Roman numerals to add and subtract problems on paper — it is not easy! He will quickly see one reason why abacuses made more sense for the Romans than paper methods.

♦ **Reinforce addition and subtraction with everyday applications.** Applying addition and subtraction to everyday situations can teach your child to both view and use these operations as useful tools. And everyday situations involving addition or subtraction abound!

» **Shopping** — As you shop bargains or use coupons, have your child help you predict your total before heading to the checkout. Suppose you had a coupon for $5 off any purchase greater than $70. You found a skirt for $30, a shirt for $20, and a pair of pants for $25. If you bought these items, would you be spending enough money to use your coupon? If so, what would your total be?

You may also want to have your child practice calculating the change you should receive. Say you bought $17.99 worth of groceries with a $20 bill. Ask your child how much change you should get back ($20 - $17.99 = $2.01). Remind your child he is using math to help him in real life. Note: If your child has not learned to work with decimal numbers yet, round the amounts before you ask him to find the appropriate change.

» **Planning a trip (how much a trip will cost)**

» **Eating out (how much the meals will cost)**

» **Balancing a checkbook and reconciling a bank statement** — An Internet search for "printable check register free" should turn up several sites from which you can download a free checkbook template. Print the template, then give your child some sample entries like the ones below to enter. After each entry, have him compute the running balance. (The entries below will result in an ending balance of $1,375.) If your child is mature enough and proficient in decimal addition and subtraction, consider letting him help you keep your checkbook for a few months.

- 3/1 Started with opening balance of $800.
- 3/2 Paid electric company $45 with check 172.
- 3/6 Gave charity $400 with check 173.
- 3/8 Received paycheck of $1,000.
- 3/18 Withdrew $30.
- 3/19 Deposited $50 gift.

» **Computing age (how long a company has been in business, how long the luncheon meat has been in the refrigerator, how old someone is, etc.)**

» **Dealing with time problems (how long has dinner been in the oven, how long has Johnny been napping, etc.)**

60   REVEALING ARITHMETIC

◆ **Explore some of addition and subtraction's applications in different occupations.** Look at how people in various jobs employ addition. The homemaker adding numbers in her checkbook and the chemist adding chemicals must both rely on the consistent way God holds things together in order to complete their tasks. Regardless of what occupation God calls your child to one day, he will find addition and subtraction useful. Teach him now to both recognize these concepts' dependency on God and to learn to think through real-life problems! Below are a few examples you could use as a launching pad.

> » **Zoo Keeper** — Have your child pretend to be a zookeeper and add up the amount of grain he would have to purchase to feed a zebra or how much it would cost to feed a snake.
>
> » **Veterinarian** — Have your child add up how much to charge someone for doing various procedures to a variety of different pets, then subtract an insurance discount.
>
> » **Librarian** — Have your child add up the price of books from catalogues.
>
> » **Nurse or Doctor** — Have your child practice adding the calories he eats each day and compare them to the recommended calories with subtraction.
>
> » **Pilot** — Have your child look into how pilots find the weight and balance of a plane before take off, then find pretend weight and balances for various scenarios (different weight of fuel, passengers, cargo, etc.).

The possibilities are endless! Consult your library or the Internet for ideas on the applications of addition and subtraction in the industries of interest to your child. If your child is old enough, have him do this legwork himself and write his own addition and subtraction problems based on what he finds!

If your child does not quite know enough math to solve actual problems from an industry of interest, try using simplified numbers and scenarios. Suppose the grain only cost $10 a day, the veterinarian only charges $5 for supplies plus $20 for the visit, and the books in the librarian's catalogue only cost $5 each.

◆ **Have your child start his own business or think through possible businesses he could start.** Starting his own business (mowing lawns, raking leaves, sewing pot holders, pet sitting, etc.) will give your child an opportunity to use both addition and subtraction, as well as other math concepts, in a practical way. If your child is not ready to start an actual business, you could have him think up a few possible businesses and calculate how much it would cost to begin by adding up the necessary supplies.

PARTING NOTE

God's ability to keep 5 plus 4 equaling 9, 20 plus 6 equaling 26, and 21 minus 7 equaling 14, reminds us of His ability to do exactly what He says He will.

> *Now unto him that is able to keep you from falling, and to present you faultless before the presence of his glory with exceeding joy, To the only wise God our Saviour, be glory and majesty, dominion and power, both now and ever. Amen.*
>
> JUDE 1:24-25 (KJV)

God is able to save us. He is able to bring us safely to Heaven. He is able to keep those who have trusted Him and bring them safely to heaven. Addition and subtraction continually remind us of His ability to save.

Addition and subtraction also remind us of God's ability to punish sin. Those trusting themselves or their own good works should tremble when they look at addition and subtraction, knowing that the God who holds all things together is able to destroy those who have not placed their faith in Him.

> *There is one lawgiver, who is able to save and to destroy: who art thou that judgest another?*
>
> JAMES 4:12 (KJV)

Multiplication:
Foundational Concept

We often need to add the same numbers over and over again. For instance, if we made $9 dollars an hour and wanted to find out how much we would make in 8 hours, we would have to add $9 eight times — $9 + $9 + $9 + $9 + $9 + $9 + $9 + $9.

Imagine that every day you had to do the same tedious task over and over again. You would look for a faster way to get the task done, wouldn't you?

Multiplication is a faster way of looking at repeated additions. Rather than looking at $9 + $9 + $9 + $9 + $9 + $9 + $9 + $9 as addition, we could think of it in terms of taking $9 eight times. It is helpful to think of this as 8 groups of $9.

Rather than writing $9 + $9 + $9 + $9 + $9 + $9 + $9 + $9, we could write 8 × $9 or 8($9) or 8 · $9. These are all agreed upon shortcuts to represent 8 groups of $9, or $9 + $9 + $9 + $9 + $9 + $9 + $9 + $9. When we view repeated additions in terms of how many times we are adding the number (in this case, 8 times), we call it multiplication. Just as there have been many symbols for addition and subtraction, there have been many symbols for multiplication.

Ready for the amazing part of multiplication? $9 + $9 + $9 + $9 + $9 + $9 + $9 + $9, or 8 × $9, *consistently* equals 72. Quantities combine so consistently we can memorize how they will multiply.

Why do quantities combine so consistently? As we have seen in the last few chapters, quantities operate consistently because Jesus upholds them consistently by the word of His power (Hebrews 1:3). Our very ability to use multiplication rests on God's power.

To better understand this important truth, let us explore together the multiplication table and the properties of multiplication. Both of these aspects of multiplication help us clearly see multiplication's reliance upon God.

Multiplication Table

1	2	3	4	5	6	7	8	9
2	4	6	8	10	12	14	16	18
3	6	9	12	15	18	21	24	27
4	8	12	16	20	24	28	32	36
5	10	15	20	25	30	35	40	45
6	12	18	24	30	36	42	48	54
7	14	21	28	35	42	49	56	63
8	16	24	32	40	48	56	64	72
9	18	27	36	45	54	63	72	81

The Multiplication Table — A Glimpse of God's Power

As a child, I used to be fascinated by the multiplication table. It seemed mind-boggling to me that this table could so accurately help me find the answer to a multiplication problem. Was there something magical about writing numbers in a table?

Years later, a look at history "revealed" the multiplication table for me, showing me how it worked and how, above all, it testified to God's power. The multiplication table is very similar to "Napier's rods" or "Napier's bones," a device developed by the Scottish Protestant John Napier back in 1617. Join me in taking a quick look at this device.

Napier's rods consisted of rods made out of ivory (which gave them their name — some sources say the rods looked like bones, others that some versions might have been made out of bones) on which multiplication facts were engraved. For example, different repetitions of 2 would be engraved on one rod. The first box of this rod would record what we would get if we took 2 one time (2 or 1 × 2), the second if we took 2 two times (2 + 2 or 2 × 2), the third if we took 2 three times (2 + 2 + 2 or 3 × 2), and so forth. Diagonal lines in each box separated the ten's column from the one's column.

When a user needed to find what 2 + 2 + 2, or 3 × 2, equalled, all he would need to do is look in the third box of the 2 rod!

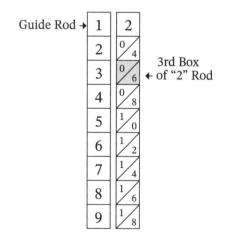

In this way, Napier's rods recorded how repeated groups of 1, 2, 3, 4, etc. added together. Rather than hand counting 2 + 2 + 2, users had only to look at the answer on the rod.

These rods saved people a lot of time, especially before the days of calculators. Using them, men could quickly multiply numbers, even large numbers!

Notice how Napier's rods are quite similar to the multiplication table. The multiplication table is essentially the same thing, only all the strips have been combined and the diagonal lines separating the ten's places from the one's places have been removed. The multiplication table works the same basic way Napier's rods does! Both organize the results of various multiplications in an easy-to-reference manner.

Do you catch the glimpse of God's power here? The multiplication table would be absolutely pointless unless repeated groups of objects added together *so* consistently one could indeed find the answer once, record it in a table, and never have to find the answer again! We can only use multiplication tables and multiplication itself because God is holding the universe together in a consistent manner.

In a sense, the multiplication table is a record of God's power and abilities! It records the consistent way He causes objects to multiply. If 9 groups of 4 did not consistently equal 36, it would be pointless to write this number down in a table to use in the future. But *because* God is all-powerful and faithful, we can use devices like the multiplication table.

Properties — God Is God, and We Are Not!

Both addition and multiplication have certain properties, or characteristics, that hold true in every situation, including those listed below.

**Commutative Property
of Addition and Multiplication**

$1 + 2 = 2 + 1$
$1 \times 2 = 2 \times 1$

**Associative Property
of Addition and Multiplication**

$(1 + 2) + 3 = 1 + (2 + 3)$
$(1 \times 2) \times 3 = 1 \times (2 \times 3)$

Distributive Property

$2(3 + 2) = (2 \times 3) + (2 \times 2)$

These properties, as well as other properties, basically state the obvious. Because of the consistent way God created the universe, it does not matter in what order we add numbers, nor how we group them, nor if we distribute them differently — the answer will remain the same. Why? Because that is how God designed it!

Although we could change how we record multiplication, we could never change the fact that 1 group of 2 objects will equal the same as 2 groups of 1 object. Properties are simply ways of stating truths about how God put this universe together.

We cannot change properties because God is who He is, and He will go on governing this universe the way He has decided, whether or not we like it. Our culture continues to reject God and His authority and seeks to reinterpret His Word. Yet, as these properties remind us, God is God and we are not.

A Useful Tool

While we will explore more of multiplication's applications in a few chapters, I wanted to mention multiplication's usefulness here. We do not memorize multiplication tables for fun — we memorize them because multiplication saves us time in hundreds of settings throughout our lives. Multiplication can help us figure out how much we would make selling an item (20 sales at $3 profit a sale), the quantity of items necessary for a project (how many pencils are needed in order to provide 5 pencils per guest for 20 guests), the total spent per year (how much is spent each year on trash pick up if $18 is spent each month), and more.

Conclusion

Quantities multiply consistently enough for us to memorize, testifying to God's power, ability, and faithfulness. We can also observe characteristics of multiplication, call those characteristics properties, and rely on those properties to work because an unchanging God governs this universe. Multiplication's very existence points us to God — and multiplication proves a very useful tool.

TEACHING SUGGESTIONS AND IDEAS

Objective: *To help your child view multiplication as a useful shortcut for expressing the consistent way God causes objects to combine. To help your child master basic multiplication facts and apply them in real-life settings.*

Specific Points to Communicate:

- *Multiplication is a shortcut for repeated addition.*
- *The multiplication table records the consistent way God causes quantities to multiply.*
- *Properties are ways of describing a consistency God determined and holds in place. Also note that the purpose of looking at different methods isn't to master multiple methods — students should just pick one to learn — but rather to help students see math methods as one* way of describing the consistencies of God's creation.

As you teach multiplication, demonstrate how multiplication facts help us with our God-given tasks. Look for real-life opportunities for your child to count groups of objects — both while introducing and reviewing multiplication. Real-life applications go a long way in building a biblical perspective of math as a tool to help us record quantities.

Example

Below is a multiplication presentation from a first-grade curriculum. The page from which these were taken is highly graphical and has pictures demonstrating the idea of multiplication.

> *This is **multiplication**. It means **putting together equal groups**.*
>
> *We write the number sentence:*
>
> *4 × 2 = 8*
>
> ***Multiply** 4 and 2. The answer is 8.*
>
> *There are 4 equal groups. There are 2 blocks in each group. There are 8 blocks altogether.*[41]

How could you reveal this presentation, helping your child see the "big picture"? You could start by rewording some sentences. For example, you could say something like this:

> ***Multiplication** is a fancy name for this process of **putting together equal groups**. One way to express the multiplication shown here is to write the number sentence 4 × 2 = 8. There are many other ways to express multiplication, but this is the one we will be using in this book. Because of God's power in holding things together, when we **multiply** 4 and 2, we consistently get 8...."*

You could also describe multiplication as a shortcut for adding repeated numbers quickly, illustrating this point with a real-life example. For example, you could have your child pretend he mowed 7 lawns a week and got paid $5 a lawn. In one week, he would make $5 + $5 + $5 + $5 + $5 + $5 + $5. That is not easy to write out! After making this point, you could present multiplication as a shortcut to abbreviate problems like this to 7 × $5 (7 groups of $5). You could actually demonstrate this with play (or real) money to make the point visually.

Do you see how you have now introduced multiplication as a shortcut dependent on God rather than as an independent method? You have also helped your child see the usefulness of multiplication by connecting it with a real-life situation rather than only pictures or manipulatives. The next section offers some additional ideas to convey these points.

Ideas

◆ **Look for multiplication applications in your child's life.** While many multiplication applications require knowing how to multiply larger numbers, there are many simple ways to have your child apply multiplication right from the

---

41. Primary Mathematics Project Team, *Primary Mathematics 1B*, Textbook, 47.

beginning. Note: You may need to round dollars to the nearest whole dollar to avoid working with decimals at this point.

If your child makes $5 a week walking neighborhood dogs, have him compute how much he will make in 5 weeks (5 × $5).

If enjoying a bag of bite-size candies, have your child practice grouping them and adding the groups. Point out the difficulty in reading equations like 2 + 2 + 2 + 2 + 2. Multiplication makes these type equations much easier to read!

If at the store buying party supplies, have your child figure out how many packages of party hats, party favors, etc., you need to purchase. He will need to multiply the number of items in a package by the number of packages to find the total items. Remind him that because of God's faithfulness and power in holding all things together, he can memorize what 5 × 5, 4 × 3, etc., equals.

You might be surprised where you will find simple opportunities to apply multiplication. One day while standing in line to talk with someone about tires for my car, I watched another family in line encourage their young children to use math to find the number of tires hanging on the wall. This could be done by counting the number in each row and the number of rows, then multiplying to find the total.

◆ **Explore the multiplication table and "Napier's rods" with your child.** Your child might enjoy making a paper version of Napier's rods. Back in the worksheet section, you will find a paper template titled "Multiplication: Foundational Concept — Napier's Rods." Make 3 copies and have your child cut around the dark lines so he has individual strips. Once the strips are cut, have your child begin filling them in with multiplication facts. The final strips should look like those in figure 5. Notice how the top box (the one with the large number) in the "1" strip represents 1 × 1, the second box 1 × 2, and so forth. The diagonal lines in each box separate the ten's column from the one's column.

I would suggest starting with the "1" strip, then doing the "2" strip, and on up in order. You may also wish to grab some manipulatives such as dry noodles or beans and have your child find each fact using the manipulatives before he writes it on the strip. For example, have him build 2 groups of 2, otherwise known as 2 × 2 or 2 + 2, using the manipulatives. Explain that since God consistently holds all things together, we can save ourselves from having to count objects next time we need to add 2 groups of 2. Instead, we are going to write 4 on the second box of our 2 strip to record what 2 groups of 2, or what we call 2 × 2, equals.

Figure 5: Napier's Rods

Once you have finished the strips, have your child practice solving problems with his rods. The "Multiplication: Foundational Concept — Using Napier's Rods" worksheet on page 174 has two sample problems to get you started.

Eventually, you can have your child begin memorizing these multiplication facts so he will be able to multiply without Napier's rods. Again, emphasize God's power and faithfulness, which make memorizing these facts possible.

At some point, have your child put all his strips next to each other and compare them to a multiplication table. Both record the consistent way repeated groups of objects add together — a consistency God upholds by His power.

◆ **Explore the life of John Napier.** You should be able to find ample information on John Napier's life online. The Lord gifted Napier with the ability to invent many contrivances to help make math computations easier, including Napier's rods. He is probably most famous for developing logarithms, a method of simplifying certain advanced calculations. While studying John Napier, you can discuss both the benefits of his practical applications of math as well as the harm of his non-biblical views and actions. Appendix A offers some questions to guide your study.

◆ **Use an abacus to help reinforce that multiplication is just adding the same value repeatedly.** Appendix C explains how to use an abacus in multiplication. When teaching skip counting, you can also make an abacus stringing columns by 2s, 3s, etc. Instructions are also in the appendix.

PARTING NOTE

As you teach multiplication, ponder God's power. God spoke, and the earth came into existence. God continues to uphold everything by His powerful Word in such a consistent way we can memorize what we will get if we take a specific number of objects a specific number of times. Is anything too hard for God? Is any situation too difficult for Him to handle? Look for teachable moments where you can remind your child of God's power. God is powerful enough to give each one of us the strength to walk victoriously every moment, if we will only let Him.

Division:
Foundational Concept

Suppose you had 12 cookies you wanted to share evenly amongst 3 people. You would end up giving each person 4 cookies.

In math, we would call this division. *Division* is a name we use to describe taking a quantity and separating it into "parts, areas, or groups."[42]

Guess what? Because God upholds all creation together consistently, if you divide 12 objects into 3 piles, you will consistently end up with 4 objects in each pile! God's consistency makes it possible to find how much we would have after dividing objects without actually touching the objects we are dividing. Just as we did with addition, subtraction, and multiplication, we can memorize our division facts and be confident they will work because of God's faithfulness.

Not only do biblical principles give us a framework for understanding why we can memorize how quantities divide, they also give us a foundation for approaching terms and signs used in division, as well as the order of operations followed in math. We can look at *every* aspect of math from either a dependent heart (a heart that recognizes that without God we can do nothing) or an independent heart.

Since we have already talked extensively in the preceding chapters about how God's power and faithfulness make math operations possible, we will take this chapter to look at some less-obvious aspects of division.

Division Terms — Names in Disguise

Terms like *dividend*, *divisor*, and *quotient* sure seem independent and neutral, don't they? But they are not!

Right from the very beginning of time, man has been exploring and naming God's creation. Back in the Garden of Eden, Adam named the animals (Genesis 2:20).

Names help us communicate. Terms like *dividend*, *divisor*, and *quotient* are names we use to refer to the different numbers in a division problem — names we are able to use because God gave us the gift of communication!

Dividend refers to the starting number — the initial quantity we need to divide into smaller parts. *Divisor* refers to the number of parts we need to divide the initial quantity into. And *quotient* refers to the ending quantity we will have in each part.

---

42. *Merriam-Webster Online Dictionary*, s.v. "divide" (2009), http://www.merriam-webster.com/dictionary/divide (accessed April 1, 2009).

These terms allow us to refer to a specific part in a division equation with a single word instead of a lengthy explanation. Without terms, math textbooks would be overwhelmingly lengthy and difficult to read.

Symbols — Useful Shortcuts

Figure 6 shows some of the different ways men have recorded division problems throughout the years. Some appear quite different from what we use in America today! Even today, we record division different ways (÷, ⌐ , a fraction line).

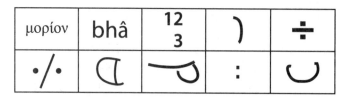

Figure 6: Different Ways to Record Division[43]

The existence of different symbols illustrates an important truth: the division sign is not an independent fact, but rather a useful shorthand for describing the consistent way God causes objects to divide!

Order of Operations — Conventions

We use a lot of conventions in math. Conventions, like grammar rules, are agreed-upon protocols to aid in clearly communicating with others.

Consider equations like 12 ÷ 3 + 3. Did we mean to divide 12 by 3, and then add 3, in which case we would get 7, or did we mean to add the 3s and then divide 12 by the total, in which case we would get 2?

To avoid this sort of vagueness, we have adopted certain conventions about the order of operations. One convention is to always solve multiplication and division before addition and subtraction. Thus, in 12 ÷ 3 + 3, 12 would be divided by 3 first. If we want to change this order or show that different numbers group together, we need to use other symbols, such as parenthesis, to show it (i.e., 12 ÷ (3 + 3) would mean to add 3 + 3, and then divide 12 by that sum — that is, by 6).

Because conventions are typically presented in the same factual way as division facts, it is easy to equate our symbols and conventions with the consistent principles God created and upholds. When we do, we can end up looking at math as a man-made system rather than as a way of describing consistencies God created and upholds.

---

43. Cajori, *History of Mathematical Notations*, 1:268–271.
    Top row, from left to right: Diophantus; Bakhshālī arithmetic; common Hindu method (using modern notation); several mathematicians used this symbol; American/English symbol used today (although many mathematicians have actually used it to represent subtraction!).
    Second row, from left to right: A variation on our current symbol used in a Buenos Aires text; symbol used by Gallimard; symbol used by Da Cunha; symbol used by Leibniz; another symbol used by Leibniz. Note: This list is by no means exhaustive!

It is important to approach conventions as conventions — as ways of describing real-life consistencies rather than as truths in themselves. Much as we could use different grammar rules (and do, in other languages), we could use different conventions. We could adopt a convention to add and subtract before multiplying and dividing. Conventions are adopted rules to aid in communication, not self-existent facts.

Conclusion

Biblical principles give us a basis for understanding each aspect of division, from division itself to division terms and symbols. Division works outside a textbook because God created and sustains an orderly universe. Terms, symbols, and other conventions all help us clearly express division; we are able to come up with names, symbols, and conventions like this because God gave us the ability to communicate.

TEACHING SUGGESTIONS AND IDEAS

Objective: *To equip your child for using division in his own life by helping him memorize his division facts. To continue to build a biblical understanding of math and why it works.*

Specific Points to Communicate:
- *Division describes a real-life principle God created and sustains.*
- *Terms, symbols, and other conventions aid in clearly expressing division.*

As you teach division, continue to build on the foundation you have already been laying. Introduce and reinforce each new aspect of division as yet another "tool" to record the consistencies around us.

I am not suggesting saying over and over again, "God is faithful, so we can record this as...." Although repeating this truth frequently might be a good idea, sometimes you may simply let this overarching perspective impact the way you teach the various aspects of division.

For example, when presenting division terms like dividend/divisor/quotient, most textbooks state something like, "a dividend/divisor/quotient is...." This wording can easily confuse the child into thinking dividends/divisors/quotients are actually some sort of fact instead of names to refer to a specific part in a division equation with a single word instead of a lengthy explanation. Try rewording as we did earlier in the chapter.

For another example, most textbooks present the division signs used today as *the* ways to represent division. Yet these symbols in themselves mean nothing — they are just symbols useful in recording the consistent way God holds the law of division together.

When presenting the division sign, you could show your child some different symbols. You have now reinforced the symbol we use as one way to represent the consistent way God causes objects to divide instead of as some sort of independent, mysterious fact.

Example

As with multiplication, most division presentations focus on presenting division facts without explaining why those facts work or making it clear our method of expressing division is just one convention. The omission of these important elements can easily leave the child viewing division as a fact independent from God instead of dependent on Him. Consider the example below.

Divide 8 mangoes into 2 equal groups. There are 4 mangoes in each group.

We write:

$8 \div 2 = 4$

Divide 8 by 2. The answer is 4.

*This is **division**. We divide to find the number in each group.*[44]

Wording like "this is division" subtly mixes up our convention for recording division with the real-life consistency. Students reading this lesson are not left with a clear understanding of why division works or of its purpose.

Let us take a look at a few things you could do to change up this lesson so as not to leave your child viewing division as an independent fact.

After looking at how the mangos in the picture divide, you could explain why division works. You might say something like this:

Guess what? Because God holds all of creation together so consistently, if you take 8 objects and put them into 2 equal piles, you will always end up with 4 objects in each pile! God's faithfulness makes it possible for us to do division without actually touching the objects we are dividing. Just as we did with multiplication, we can memorize our division facts and be confident they will always work because of God's faithfulness.[45]

When introducing how to write division, you could present a few different ways of writing division. Of course, you would not have your child learn them all at once. After presenting the options, you can focus on the one method you plan to teach first.

Let's rewrite what we have been doing with these mangos using symbols instead of words. Symbols help us clearly communicate and save us lots of time and energy! Below are several different commonly used ways to represent division.

$$8 \div 2 \qquad 2\overline{)8} \qquad \frac{8}{2}$$

Do you see how these changes and additions can help your child view division as dependent on God rather than independent from Him? And you do not have to stop there! You could also demonstrate division with more manipulatives, thereby emphasizing the real-life consistency division facts represent. Or you could look for opportunities for your

---

44. Primary Mathematics Project Team, *Primary Mathematics 2A*, Textbook, 81.
45. Sample presentation adapted from Loop, *Beyond Numbers*, 42.

child to apply division in his own life. You could talk with your child about God's power, reminding Him God is able to help us live victoriously.

The Bible tells us Christ Jesus is "upholding all things by the word of His power" (Hebrews 1:3). God's power is amazing! His Word is powerful enough to govern the way all the objects in the entire universe operate! It is powerful enough to make objects consistently divide the same way.

Never forget we serve a powerful God. Sometimes we can feel discouraged by ourselves. But God is powerful enough to save us and to help us live victoriously — if we will just let Him. Let us join Paul in asking God to enlighten our understanding to grasp His power.

> *The eyes of your understanding being enlightened; that ye may know what is the hope of his calling, and what the riches of the glory of his inheritance in the saints, And what is the exceeding greatness of his power to us-ward who believe, according to the working of his mighty power, Which he wrought in Christ, when he raised him from the dead, and set him at his own right hand in the heavenly places,*
>
> `EPHESIANS 1:18-20 (KJV)

When it comes time to present terms and order of operations, present them as names and conventions used to communicate more clearly. This can usually be done by slightly rewording your textbook definitions to make it clear the terms are ways of referring to numbers in a division problem and the order is an adopted convention to avoid confusion.

As you approach division with the understanding of its dependency on God, you will be amazed at how many fun ways there are to "reveal" it. See the section below for a few more ideas.

Ideas

- ◆ **Look for simple opportunities for your child to apply division.** Real-life examples can help you show division's true purpose. We frequently need to take a number and divide it, so finding examples should not be hard!

    Say you spend $40 a year for a magazine that comes 4 times each year. To find the cost per magazine, you would need to divide the cost by the number magazines received, or 40 ÷ 4, which equals $10. While $40 a year did not sound like a lot of money, $10 per magazine seems a bit steep! You may want to have your child find the cost you pay per month on gas, electricity, trash, or some other bill. (You may need to simplify the numbers for younger children.) You could also have your child divide up toy trucks, dolls, cookies, or other items with more meaning to him.

    Some other ideas would include having your child share (divide toys, food, etc.) among several people, form teams (divide the total number of people into separate teams), prepare for a party (divide the total number of party supplies by the people coming), or go on a treasure hunt (divide the spoils).

Note: Many other division applications require knowledge of long division; the ideas listed above could easily be presented without knowing long division.

- ◆ **Demonstrate division with manipulatives.** It is important for your child to view division as more than busywork — you want him to know division describes a real-life consistency God created and sustains! Demonstrating division with manipulatives (objects children can use to observe math principles concretely) connects dry facts with the real-life consistency they represent. You can turn basically anything into a manipulative — toys, writing utensils, blocks, dried beans, noodles, etc.

- ◆ **Have your child divide using Napier's rods.** Although Napier's rods were designed for multiplication, they can help a child really understand division rather than just memorize facts! See page 64 for an overview of Napier's rods, and the "Division: Foundational Concept — Napier's Rods" worksheet on page 175 for details on using Napier's rods to divide.

- ◆ **Use an abacus to help reinforce that division is just subtracting the same value repeatedly.** Appendix C explains how to use an abacus in division.

PARTING NOTE

Division, like multiplication, testifies to God's power in upholding all things. As you teach your child division, praise God and stand in awe of His might. He is holding all things together by the power of His Word!

> *Who being the brightness of his glory, and the express image of his person, and upholding all things by the word of his power, when he had by himself purged our sins, sat down on the right hand of the Majesty on high:*
>
> HEBREWS 1:3 (KJV)

Multiplication:
Multi-Digit Operations

Just as we could not possibly memorize every number we will need to add or subtract in our lives, we cannot memorize every number we will need to multiply. We need a method for multiplying numbers we do not have memorized.

Men, using the ability God gave them, have developed methods to solve multiplication problems. The equation shown expresses the method most of us have come to associate with multiplication, which we will refer to as the traditional method.

$$\begin{array}{r} \overset{1}{1}2 \\ \times\ 9 \\ \hline 108 \end{array}$$

Most textbooks spend a great deal of time teaching the mechanics of this method. After all, it is important to know how to multiply multi-digit numbers.

But if we have only learned to manipulate numbers on a piece of paper, we have missed out. We have missed really understanding how this algorithm describes God's creation. We have missed learning to use multiplication in the tasks God has given us to do.

Let us "reveal" multiplication methods together by taking a look at how multiplication methods describe reality and by exploring some different methods and the purpose they serve.

Methods for Describing Reality

Multiplication methods organize large problems to which we do not know the answer into a number of smaller problems to which we do know the answer. For instance, suppose we needed to multiply 12×9 but had only memorized our multiplication facts through 10. We could think of 12×9 as the sum of 10×9 and 2×9, then add the answers together.

- » $10 \times 9 = 90$
- » $2 \times 9 = 18$
- » $12 \times 9 = 90 + 18 = 108$

Our traditional multiplication method, or algorithm, lets us perform these same steps without having to think about them. Other multiplication methods break problems into different steps or offer different ways to automate the steps.

To better see how multiplication methods break up large problems, we will take a look at two different multiplication methods: the gelosia method and the traditional method.

Here are some basic steps to find 9 × 12 using the gelosia method:

1. Draw several squares and divide those squares into triangles.

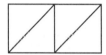

2. Write the two numbers being multiplied above and to the right of the squares.

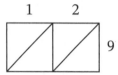

3. Write the answer to 9 × 2 in the first square.

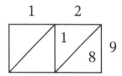

4. Write the answer to 9 × 1 in the second square.

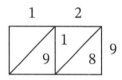

5. Add the numbers in the little triangles diagonally from right to left. In the first diagonal, we have 8, so we write 8 underneath the first triangle. The triangles with the 1 and the 9 are both within the same diagonal, so we add them together and get 10. This gives us our answer: 108. Note: The triangles are shaded in the picture to illustrate which numbers were added together. We wrote the 1 along the left to remind us it was in the hundred's place. We then read the number from left to right to get 108.

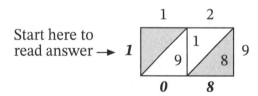

Now let us take a closer look at what happened here. In the gelosia method, the squares and triangles separated the ones, tens, and hundreds. When we multiplied 9 × 2, we wrote the 1 part of 18 in the upper triangle because that 1 really represented 1 group of 10. Likewise, when we multiplied 9 × 1, we wrote the 9 in the ten's section of the second square because the 9 really represented 9 tens (the 1 in 12 stands for 1 ten, so when we multiplied 9 × 1, we were

78   REVEALING ARITHMETIC

really multiplying 9 × 1 ten). When we added up the triangles diagonally, we were really adding up the ones and the tens to find the total answer.

Notice how the gelosia method is a simple way to keep track of place value while we multiplied. Rather than thinking "9 × 2" and "9 × 10," we thought of "9 × 2" and "9 × 1." But the method made sure we wrote the answer to 9 × 1 in the ten's place, since that 1 really stood for 1 ten.

A very similar thing happens in the method with which most of us are familiar. Rather than jumping right to the final way we multiply, let us develop the method step by step by breaking apart the equation 12 × 9 into smaller equations (9 × 2 and 9 × 10) and solving.

$$\begin{array}{r} 1\,2 \\ \times\ \ 9 \\ \hline 1\,8 \\ +9\,0 \\ \hline 1\,0\,8 \end{array}$$ (9 × 2) One's column
(9 × 10) Ten's column

Now, when we have a task to do over and over again, it is always good to find ways to save steps! There is a way to save steps when we multiply on paper. Instead of writing 18 underneath the line, we could write only the 8 under the line, and put the 1 up on top of the 1 in 12. This reminds us we have 1 ten we will need to add to the ten's column.

$$\begin{array}{r} ^1 \\ 1\,2 \\ \times\ \ 9 \\ \hline 8 \end{array}$$

Now we can work the ten's column. 9 times 1 ten equals 9 tens, or 90, plus the 1 ten we need to add gets us to 10 tens, or 100. If we add 100 and 8, we will get the answer, 108.

$$\begin{array}{r} ^1 \\ 1\,2 \\ \times\ \ 9 \\ \hline 8 \\ +\mathbf{1\,0\,0} \\ \hline \mathbf{1\,0\,8} \end{array}$$

Now there is still a way to save even more steps. Instead of writing the 100 underneath the 8 and adding it, we could have written 10 next to the 8. Since the 10 is in the ten's column, it stands for 10 groups of 10, or 100.

$$\begin{array}{r} ^1 \\ 1\,2 \\ \times\ \ 9 \\ \hline \mathbf{1\,0}8 \end{array}$$

The rules we follow when we multiply keep track of place value, thereby allowing us to break multi-digit multiplication problems we do not have memorized into a series of smaller problems we do have memorized. A multiplication method will only work if it accurately describes the way God causes objects to multiply. If God were not faithfully holding all things together, reducing multiplication to a method would be impossible.

Many Different Methods

Since multiplication methods describe a real-life consistency, we would expect different people to effectively use different methods. And they do!

Back when written arithmetic methods were first becoming popular in Europe, people experimented extensively with different multiplication methods. I have been continually amazed to discover yet another method or variation on a method. Sometimes, too, the same method had multiple names. The gelosia method, for example, was also called the "quadrilateral, the square, or the method of the cells, and to the Arabs after the 12th century by such names as the method of the sieve or method of the net."[46] People often named a method after whatever they thought it resembled, and sometimes different people chose different names.

Even today, many people use different multiplication methods, some of which are quite different from the typical one taught!

Figure 7 shows just a few of the various methods used throughout history — notice some of them differ only slightly from the method typically taught in math textbooks, and others differ drastically. Note: Appendix D includes an explanation of each of these methods not already covered.

The many different multiplication methods out there remind us that, far from being man-made systems, multiplication methods describe a real-life consistency. Why else would so many different people find methods to arrive at the same answers? Each and every one of these methods ultimately rests on God's faithfulness in holding all things together.

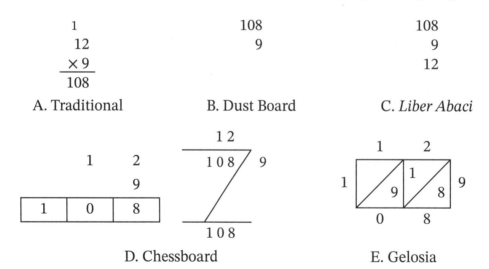

---

46. Smith, *History of Mathematics*, 2:114–115. Smith offers several footnoted sources for this statement.

1	12		12	12\9
2	24	(12 + 12)	× 9	98
4	48	(24 + 24)	008	1
8	96	(48 + 48)	108	108
9	108	(96 + 12)		

F. Duplication   G. *Speed Mathematics Simplified*   H. Juan Diez

Figure 7: Different Multiplication Methods

Multiplication — A Useful Tool

Since multiplication describes a consistent way God governs all things, we should expect multiplication to prove useful in both exploring the world He created and completing the tasks He gives us. And it does!

Have you ever wondered how we know the distance light travels in a year? We cannot physically measure the distance light travels in a year — no measuring tape can extend far enough. We *can*, however, measure the distance light travels in a second. Using multiplication, we can then compute how far light travels in a year!

The section below shows how, starting with the knowledge light travels approximately 186,282 miles a second, we can use multiplication to calculate the distance light travels in a year.[47]

- » Seconds in an hour = 60 (seconds in a minute) × 60 (minutes in an hour) = 3,600
- » Seconds in a day = 3,600 (seconds in an hour) × 24 (hours in a day) = 86,400
- » Seconds in a year = 86,400 (seconds in a day) × 365.25 (days in a year) = 31,557,600
- » Distance light travels in a year = 31,557,600 (seconds in a year) × 186,282 miles (distance light travels in a second) = 5,878,612,843,200 miles

We use light-years as a measurement unit to express the vast distances of space. But just because a star is 30,000 light-years from earth does not mean the light took 30,000 years to get here! It only means the star is an incredible distance away! As we just saw, a light-year is just a useful measurement unit based on *the distance we observe light travel in a second here on earth.* Light may very well travel differently in the deep recesses of space, or God may have used some other way to get the light here much faster than normal.[48]

---

47. This example is included here to give you a better understanding of multiplication's purpose. You would not want to share this with your child unless he has dealt with multiple-digit multiplication in the past and is doing this chapter as a review.
48. A question that often arises when discussing cosmic distances (of stars, galaxies, or other celestial objects) from a biblical perspective is, "How can light from distant stars be seen in a 'young' universe?"
    The simple answer is that, in fact, we do not really know. But neither do secular scientists know how light travels. In fact, light-travel time is also one of the big problems for the old-earth Big Bang theory (http://www.answersingenesis.org/creation/v25/i4/lighttravel.asp). The fact is no one fully understands how light, gravity, and other physical forces operate on a cosmic scale.
    There are many ways to explain distant starlight. Dr. Russell Humphreys lays out a complete young-earth cosmology, based of the work of Albert Einstein, that neatly solves the distant starlight problem, as well as many other problems secular scientists have been unable to explain (http://creation.com/images/pdfs/cabook/chapter5.pdf). His work has

Not only does multiplication aid in measuring the distance light travels in a year, it also helps us appreciate God's creation more fully. When we read in a textbook that the *closest* star to earth, Proxima Centauri, is about 4.3 light-years away, it does not sound that far away, does it? But if we use multiplication to convert 4.3 light-years to miles, we will gain a better appreciation for the magnitude of God's universe — even this close star is hardly close!

$$4.3 \times 5{,}878{,}612{,}843{,}200 \text{ miles} \approx 25{,}000{,}000{,}000{,}000 \text{ miles}$$

Imagine driving from Massachusetts to California. According to MapQuest®, the trip is about 3,000 miles and would take more than 40 hours to drive. Using math, we can see this trip is only about 1.2 ten-billionth ($\frac{1.2}{10{,}000{,}000{,}000}$) of the distance to Proxima Centauri!

We can hardly even understand distances like 25,000,000,000,000 miles, and most stars are much, much further away! How much more incredible the God who stretched out the heavens!

> *I have made the earth, and created man upon it: I, even my hands, have stretched out the heavens, and all their host have I commanded.*
>
> ISAIAH 45:12 (KJV)

> *Behold, the nations are as a drop of a bucket, and are counted as the small dust of the balance: behold, he taketh up the isles as a very little thing.*
>
> ISAIAH 40:15 (KJV)

How much greater God is than us! Yet this great God knows and cares for each of His children personally.

> *I am the good shepherd, and know my sheep, and am known of mine.*
>
> JOHN 10:14 (KJV)

By helping us grasp the magnitude of distances like 4.3 light-years, multiplication gives us a better appreciation for the greatness and incomprehensibleness of God, the One who stretched out the distances we cannot even comprehend.

Multiplication also aids in a variety of tasks we face throughout our lives. It has a way of proving useful in the most unlikely places. One spring, our church held a Passover celebration. Everyone attending needed to bring a dish to share. We signed up to bring beef brisket, but we had no idea how much brisket to purchase! Fortunately, the man at the meat counter knew we needed approximately 4 ounces a person. Since we wanted to bring

---

sparked even more research, and there are now several creationist cosmologies (including one by Dr. John Hartnett, author of *Starlight, Time, and the New Physics*) based on "gravitational time dilation." Dr. Jason Lisle, author of *Taking Back Astronomy*, has proposed the "Alternate Synchronization Model," suggesting the universe is divided into "time zones" similar to what we have on Earth (http://www.answersingenesis.org/tj/v15/i1/starlight.asp).

While we do not know exactly how God did it, we do know God created the stars to give light upon the earth. Getting the light here was no problem for our God! "And God set them in the firmament of the heaven to give light upon the earth." Genesis 1:17 (KJV)

For more information on this important topic, please check out the resources on www.answersingenesis.org.

(Thanks to Zak Klein for his help with this footnote.)

enough for 20 people, we needed about 20 × 4, or 80 ounces, which converts to 5 pounds since there are 16 ounces in a pound (80 ÷ 16 = 5). Since the brisket cost $4.99 a pound, we knew it would cost us about $24.95, since 5 × $4.99 = $24.95.

Assumptions Matter

Multiplication can only give us valid answers if we multiply the correct numbers and interpret the answers with the correct assumptions. For example, suppose we said, "There are 40 rooms taken at this hotel, and each room can sleep 2 people, so there must be 80 people staying in the hotel." While it is true that 40 × 2 does equal 80, our statement is not necessarily true. It assumes 2 people are sleeping in each room, which is not necessarily true.

This is an important point to grasp. Just because someone tells us they have mathematically proven something does not mean what they have "proven" is necessarily true. The assumptions behind math make a difference. We need to be careful not to place our faith in math itself.

To use multiplication effectively, the appropriate numbers need multiplied! Otherwise, completely correct multiplication could result in an untrue answer.

Conclusion

Multiplication can be reduced to a method because of the amazing consistency in the way God causes objects to multiply. Because multiplication describes a real-life consistency, we use it to learn about and explore God's creation, all the while seeing glimpses of God's character. When used appropriately, multiplication serves as a useful tool in the work God has given us to do.

TEACHING SUGGESTIONS AND IDEAS

Objective: *To teach your child to effectively multiply multi-digit numbers with the view of using this knowledge as a practical tool.*

Specific Points to Communicate:

- *The multiplication method being taught is one way of simplifying the process of multiplying numbers.*
- *Multiplication can only be reduced to a method because God holds all things together consistently.*
- *Multiplication helps us explore God's creation and aids us in our daily tasks.*

$$\begin{array}{r} 1 \\ 1\,2 \\ \times\ \ 9 \\ \hline 1\,0\,8 \end{array}$$

Teach multiplication methods as logical methods rather than rote rules. For example, in the equation 12 × 9, instead of saying, "We carry the 1 to the next column," show your child how the 1 represents 10 objects, so we need to put it over in the column we use to represent groups of 10; carrying is one method for moving numbers representing groups of 10 to the ten's column.

You may find it useful to demonstrate multiplication using a household manipulative (crayons, paper clips, dry beans, etc.), visually demonstrating the consistency the method records. This will both help your child understand the method, and, more importantly, teach him to view math as a tool to describe God's creation instead of as some sort of man-created system.

Later, as your child masters multiplication methods, help him apply them outside a textbook. Teach him to explore the distances to stars, organize events, evaluate options, and much, much more! Practical explorations can go a long way in preparing your child to use math effectively in whatever task or situation the Lord may one day send his way.

Example

Teaching math from a biblical worldview means so much more than *saying* over and over again, "God created and sustains a consistent universe and math records it." It is a fascinating journey of actually *showing* this and teaching the child to use each new tool while praising the Creator.

I know most multiplication presentations seem more factual than fascinating. Your child's textbook might read something like the paragraph below, and be filled with apparently meaningless practice problems.

> *Multiplying a two-digit number by a one-digit number.*
> *Step 1: Multiply 3 times 7 ones.*
> *3 sets of 7 ones = 21 ones.*
>
> *Step 2: Rename 21 ones as 2 tens and 1 one.*
> *Write a 1 in the ones place.*
> *Write a 2 in the tens box...*[49]

Although factual-sounding, these steps are important. These steps explain how to use an important tool. If we were given a special kit to build, we would pour over the instruction manual until we had mastered what we needed to do and could build the kit. Multiplication steps, likewise, help us master a useful tool.

One way you can help your child grasp this is to have him use multiplication outside a textbook. In fact, do not be afraid to break from the textbook altogether and substitute the textbook's problems with real-life situations for your child to solve. Help him learn how to use this tool rather than forcing him to spend hours mastering a method that apparently has no purpose. Children need to see where multiplication leads — how *they* can use it as a tool!

> *There can be nothing more destructive of true education than to spend long hours in the acquirement of ideas and methods which lead nowhere.*[50]
> ALFRED NORTH WHITEHEAD

---

49. Jacobs, et al., *Math 3 for Christian Schools: Teacher's Edition*, 463 (Lesson 129). Quoting from the student booklet, 257.
50. Alfred North Whitehead, *Essays in Science* (London: Rider and Company, 1948), 133. Quoted in Nickel, *Mathematics: Is God Silent?* 289.

As you teach multiplication steps, present the thinking behind those steps — show how the steps describe what happens in real life. Most textbooks try to some degree to communicate this, but the student does not always get this point. It often gets lost in his attempt to memorize the steps, solve the problems, and finish math for the day.

You can use other multiplication methods to help your child better understand the thinking behind the method he is learning. Looking at other methods can also help your child see multiplication as a method dependent on God and the consistencies He sustains rather than as a man-made or self-existent fact.

Bringing in science can also be quite beneficial. As we use math in science to explore God's creation (such as a light-year), we continually see glimpses of God's character.

Wondering how to go about doing all that? Keep reading! The section below offers some hands-on ideas to help you accomplish these goals.

Ideas

- ◆ **Explore other multiplication methods.** People continue to think of different methods to help them multiply. Appendix D offers explanations of a few methods. You may even stumble upon a method your child finds easier. Note: Please use discretion when looking at other methods, as you do not want to cause confusion by switching back and forth between methods. Also note that the purpose of looking at different methods isn't to master multiple methods — students should just pick one to learn — but rather to help students see math methods as one way of describing the consistencies of God's creation.

- ◆ **Look for ways to have your child apply multiplication in a real-life example.** See the "Multiplication: Multi-Digit Operations - Multiplication in Real Life" worksheet on page 176 for several examples. Each of these could serve as a launching pad for other similar examples. Keep your eyes open for ways you use multiplication in everyday life, and incorporate them into your math. If you love science, you might want to pull in an application or two of multiplication from science. You could also utilize some word problems from old math books. Back in the 1800s, math books teemed with practical examples. A list of some that are currently accessible free online are listed on the Book Extras page on ChristianPerspective.net. The books are all searchable, and searching inside one of the arithmetic books listed there for "multiplication" should take you to the appropriate sections.

- ◆ **Have your child evaluate options for a real-life situation.** We often use multiplication when finding costs and evaluating options. For example, suppose your family needed to spend a few months in an apartment while building a new home. You would need to decide whether to rent a furnished apartment or a non-furnished one. If you opted for a non-furnished apartment, you would need to decide whether you should rent furniture or pay to have some of your furniture delivered. To help you make your decision, you might need to find out how much each of your options would cost. You could use multiplication to help you!

The "Multiplication: Multi-Digit Operations - Apartment Rental" worksheet on page 177 gives some imaginary data regarding apartments and different options someone in the above situation might face.

Your child will encounter many situations in life where he will want to use multiplication to help him evaluate different options. Teach him how to use math to evaluate options while depending on God and seeking His wisdom rather than relying on his own. While math can help us see the most economical choice, God may have a different choice in mind.

Consider the woman who poured perfume on Jesus' head (Mark 14:3-9). To onlookers, it appeared she completely wasted very expensive perfume. Jesus' disciples wondered why she had not given the money to the poor, but Jesus was pleased by what she had done. In His eyes, "wasting" the perfume was the best decision she could have made.

- ◆ **Point out multiplication in more advanced math concepts.** As your child learns other math concepts, he will keep learning more and more uses for multiplication. Multiplication proves handy in finding the area and perimeter of objects, computing interest rates, and much more!

PARTING NOTE

Always remember, textbooks do not have to be followed as rigid rules! If your child is not grasping a concept, ask God for a different way of explaining it, if you need to hold off on presenting it for a time, or if there is something else God may be trying to show you or your child through the struggle. God makes a wonderful teacher!

> *And all thy children shall be taught of the LORD; and great shall be the peace of thy children.*
>
> ISAIAH 54:13 (KJV)

Division:
Multi-Digit Operations

```
  24
2)48
 -40
  08
  -8
   0
```

Multi-digit division methods, like multiplication methods, are useful procedures to represent on paper what happens when we divide quantities. The picture shows a division equation using the method with which most of us are familiar.

Join me as we "reveal" multi-digit division (also known as long division) together by taking a look at a variety of division methods. We will see that, much as multiplication, division methods record a real-life consistency and prove quite useful!

Different Methods — Same Consistency

Just as there are different ways to multiply a number, there are many different ways to divide a number! The Egyptians, as well as cultures using abacuses, utilized a process called repeated subtraction to divide. To divide 95 by 5, they subtracted 5 from 95 over and over again, keeping track of the number of subtractions made.

Figure 8 shows a few other division methods more similar to the traditional one with which most of us are familiar (shown on the left for comparison). The traditional one is not necessarily the best one! Just as there are different ways to do a job, there are different ways to divide.

Repeated Subtraction

```
95    60    30
 5     5     5
90    55    25
 5     5     5
85    50    20
 5     5     5
80    45    15
 5     5     5
75    40    10
 5     5     5
70    35     5
 5     5     5
65
 5
```

Each of these methods, far from being something special in itself, is a useful technique for keeping track of digits, making it possible to accurately represent how God causes quantities to divide.

```
    19
5)95           40           19
 -50           90         5)95
  45            5            4
 -45           19
   0
```
A. Traditional   B. *Liber Abaci*   C. *Speed Mathematics Simplified*

Figure 8: Other Division Methods
(See Appendix D for explanations of the methods listed.)

Methods Describe Consistency

As a child, I proudly learned long division, all the while clueless why the long division method worked. Why do we divide numbers the way we do? How does this method work? How does it lead to the correct answer?

To answer these questions, let us look at a different way to solve the equation 48 ÷ 2 using circles as an illustration. We will then see how the traditional long division method uses this same process to automate solving division problems on paper.

Rather than trying to divide 48 directly by 2, look at the 48 circles as 4 groups of 10 and 8 individual circles.

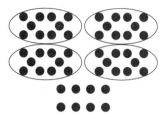

Next, divide the 4 groups of 10 by 2. Notice that just as <u>4</u> divided by <u>2</u> equals <u>2</u>, <u>4</u> piles of 10 divided by <u>2</u> equals <u>2</u> piles of 10.

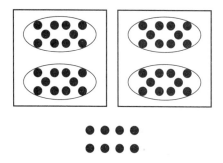

Now divide the 8 individual circles into two piles, getting 4 in each pile, and add these piles to the ones obtained from dividing 4 groups of 10, forming 2 piles of 24. 48 divided by 2 equals 24.

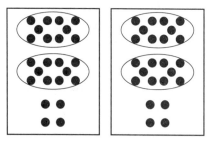

The traditional division method completes this same basic process on paper. To solve 48 ÷ 2 using it, start by writing the 48 and the 2 separated by a partial box that looks a bit like a stair. This "stair" format is a popular way to keep the numbers separated.

$$2\overline{)48}$$

Then ask, "If I were to divide 40, or 4 groups of 10, into 2 piles, how many would be in each pile?" The answer is 20, or 2 groups of ten. To represent this, put a 2 over the 4 in 48 because the 4 is in the ten's place and the 2 represents 2 groups of 10, or 20.

88 REVEALING ARITHMETIC

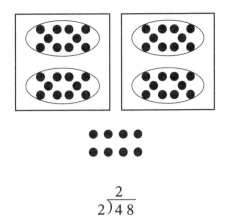

$$2 \overline{)4\,8}^{\,2}$$

Now multiply 2 × 2 (groups of 10) to find out the quantity divided so far (2 piles with 2 groups of 10 in each). The purpose of this multiplication is to verify if all the ten's groups have been divided. 2 groups of 10 times 2 equals 4 groups of 10, or 40. Write 40 underneath the 48 and draw a little line and write a 0 to show the 0 groups of 10 left over (4 groups split into 2 groups perfectly) and an 8 to show the 8 circles left to divide.

Now deal with the one's place, or the individual circles. Ask, "How many groups of 2 can be made from the 8?" and write the answer (4) up next to the 2.

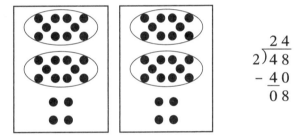

Once again, use multiplication to make sure all the circles have been divided. Multiply 2 × 4 and write an 8 underneath to represent the number of circles used to make 2 groups of 4. Minus it out of the 8 to see that all the circles have been used — 0 remain undivided. The answer to 48 ÷ 2 is 24!

$$\begin{array}{r} 2\,4 \\ 2\overline{)4\,8} \\ -\,4\,0 \\ \hline 0\,8 \\ -\,8 \\ \hline 0 \end{array}$$

Do you see how the division method and rules to divide numbers on paper break a division problem into little steps? When we divide a two-digit whole number on paper, we are really dividing the groups of 10 first, then the single items. As numbers get larger, we follow a similar process, beginning with the highest place value.

Because this method keeps track of place value for us, we can divide groups of 10 (or 100, 1,000, etc.) as easily as we would individual items. Just as 4 divided by 2 equals 2, 4 groups of 10 (or 4 tens) divided by 2 equals 2 groups of 10 (or 2 tens). We do not even necessarily have to think of the fact that we are dividing groups of 10 — the division method keeps track of all that for us. By breaking numbers down, we can use the basic division facts we have memorized to divide numbers of any size.

Division methods are good examples of how we can use the creativity and reason God gave us to observe and record His creation.

Conclusion

Division methods are useful ways to divide quantities on paper easily. Each method works because it accurately represents the real-life consistency God created and sustains.

TEACHING SUGGESTIONS AND IDEAS

Objective: *To help your child understand long division algorithms as methods to represent what happens when objects divide; to learn to divide large numbers proficiently.*

Specific Points to Communicate:

- *The division method being taught describes the consistency God placed around us; other methods also exist.*
- *Division aids in exploring God's creation and completing real-life tasks.*

As you present division methods (algorithms), encourage your child to stop and think about why each method works. You can do this by demonstrating what is actually happening using household manipulatives. Divide the manipulatives, pointing out how the algorithm, or "computational procedure,"[51] serves as a "shortcut" to describe what happens when objects divide.

Bring in real-life applications to give your child opportunities to apply math as a tool. Division is used a lot in everyday life, so finding applications should not be hard. You may find, though, that a lot of division applications involve decimal numbers. While you can simplify many of these applications by using simpler numbers or leaving the answer as a remainder or fractions, plan on reviewing division after your child learns decimals.

Example

Consider this division presentation:

> *In this lesson we will learn a pencil-and-paper method for dividing a two-digit number by a one-digit number.... For the first step we ignore the 8 and divide 7 by 3. We write the 2 above the 7. Then we multiply 2 × 3 and write "6" below the 7. We subtract and write "1."*[52]

$$\begin{array}{r} 2 \\ 3\overline{)7\,8} \\ -\,6 \\ \hline 1 \end{array}$$

Notice this presentation does not really show how the division method works at all. Why does ignoring the 8 and dividing 7 by 3 get us the correct answer? Why do we multiply 2 × 3 and write the answer underneath? Why do we need to have a pencil-and-paper method anyway?

---

51. *The American Heritage Dictionary of the English Language*, 1980 New College Edition, s.v. "algorithm."
52. Hake and Saxon, *Math 54*, 252.

Without answers to these questions, students subtly begin to blindly trust their math books rather than to use their God-given ability to think. They also miss out on really understanding how the division method taught describes the consistency God placed around us.

One way to change this presentation would be to walk through the division algorithm using a manipulative, demonstrating step-by-step how the method accurately represents the process of dividing, much as we did earlier in the chapter. Before beginning, you could introduce the need for long division.

> *Because of the consistent way God upholds all things, you have been able to memorize many division problems. But it is not possible to memorize every single division problem we might encounter. We sometimes have to divide larger numbers such as 1,785,000. You would not want to use objects to try to divide that, now would you?*
>
> *Fortunately, God gave man the ability to develop methods to help us easily solve division problems we do not have memorized by using the division facts we do have memorized. We are going to learn one of these methods today.*

You could then walk through the division problem given in the presentation (in this case, 78 ÷ 3) using physical objects (pretzels, bite-size candies, raisins, or beans) in a similar manner to what we did earlier in the chapter with circles. Your goal is to walk your child through logically deducing the method's steps so he can learn the method with understanding.

You could also mention the existence of other division methods, or share with your child some of division's applications (see the next section). If you wanted to change things up, you could start by presenting a real-life problem, and then guide your child into discovering the division algorithm for himself as a means to help him solve that problem. For example, you could have him try to calculate how many pages in a 48-page book (or some other number small enough to represent with manipulatives) he needs to read each day to finish it in a certain number of days. As he struggles to find the answer, pull out some manipulatives and show him how he could break the problem down into steps. Then show him the paper method and how it breaks problems into steps without requiring as much effort!

Ideas

- ◆ **Reinforce division's application through real-life problems.** We all use division frequently. Below are a few applications you might explore with your child.
    - » Finding the price per unit (to find the cost of an individual item sold in a package, i.e., a box of tissues from a package of 3)
    - » Dividing objects evenly among family members (such as a bag of candy)
    - » Planning (figuring out how many pages in a book need read each day to finish it on time, finding how much yarn is needed for a crocheting/knitting project, deciding how much to lay aside each month for a yearly bill)

The "Division: Multi-Digit Operations — Division in Everyday Life" worksheets on page 179 and 180 provide a few real-life word problems. There are two versions of the worksheet. Version A involves dividing smaller numbers, and Version B requires division of decimals and larger numbers. Feel free to further simplify or complicate the worksheet to fit your child's skill level. Simply replace the numbers given in a problem with numbers from an equation in your child's textbook. For example, if your child's textbook gives the problem 50 divided by 5, substitute those numbers into problem 2 on the worksheet by asking your child to find how many pages he would have to read each day of a 50-page book if he had 5 days to read it.

- ◆ **Have your child apply division by starting a small business.** Want to really show math's practicality? Have your child start a small business. He could open a lemonade stand for a day, make and sell scarves or some other piece of clothing or craft, build and market birdhouses, etc. Have him add up all the costs of making his product, then divide the total cost by the number of products he made (for a lemonade stand, the total number of cups of lemonade). He can use this information to help him set his price and determine his final earnings. A small business is a great way to let your child use division in real life, reinforcing division usefulness in completing the tasks God has given us.

- ◆ **Take a look at other division methods.** To help your older child appreciate our long division method and view it as *one* way of describing division, try showing him some other division methods. Note: Please use your judgment on whether seeing other methods (or even trying to think up his own method!) would confuse your particular child. Appendix D explains a few division methods and footnotes resources you can consult for more information. You might even want to have your child solve a few problems from his math book using one of these other methods. The repeated subtraction method is a good one to begin with, as it is quite different from the traditional one and fairly intuitive. Note that the point is not for him to master other methods, but rather to be aware that men have used their God-given abilities to approach the problem differently.

PARTING NOTE

As you teach division, ponder God's power. God keeps this universe operating so predictably we can reduce division to a method! Is anything too hard for God? Is any situation too difficult for Him to handle?

Behold, I am the LORD, *the God of all flesh: is there any thing too hard for me?*

JEREMIAH 32:27 (KJV)

Fractions:
Foundational Concept

When we speak of a fraction, we typically refer to a "part of something."[53] Yet fractions are much more than a naming system for partial quantities. Although fractions have earned a reputation for being hard to understand, they do not have to be! They are simply another way of looking at and working with the quantities around us.

Since fractions have a lot of different aspects, I have divided both this chapter and the next one into several subsections, each one of which examines a specific facet of fractions.

Revealing Fractions — What Are Fractions, Why Can We Use Them, and Why Do We Use Them?

What Are Fractions?

The word *fraction* has its root in the Latin word *frangere*, meaning "to break."[54] That might sound odd at first, but "to break" actually describes fractions quite well. Although we tend to think of fractions in terms of partial quantities, they also represent "breaking" up quantities, or division.

Say you had one orange, and you wanted to break, or divide, it among four people. Each person would end up with $\frac{1}{4}$ of the orange, right? We could look at the $\frac{1}{4}$ here two different ways: as 1 part out of 4 *or* as 1 orange ÷ 4 people.

Or say you had 2 oranges you wanted to divide among four people. Each person would end up with $\frac{2}{4}$ (or $\frac{1}{2}$) of an orange. We could look at the $\frac{2}{4}$ here two different ways: as 2 parts out of the 4 in each orange *or* as 2 oranges ÷ 4 people.

Both 1 part out of 4 and 1 ÷ 4 represent the same quantity. Likewise, both 2 parts out of 4 and 2 ÷ 4 represent the same quantity. Fractions represent both partial quantities and division.

Why Can We Use Fractions?

God gave us the ability to name and work with quantities — as well as to develop different ways to express those quantities on paper. Fractions are one tool we use to express quantities as portions of other quantities. Much as Adam used words to name animals back in the Garden of Eden, we use fractions to name partial quantities.

Fractions, like other math concepts, presuppose consistency. Fractions only have the meaning they do because 2 ÷ 4 consistently equals the same quantity as 2 parts out of 4. Ultimately, we

---

53. *The American Heritage Dictionary of the English Language*, 1980 New College Edition, s.v. "fraction."
54. Ibid.

can use and work with fractions because of the consistencies God sustains and the ability God gave us to observe and record those consistencies.

Why Do We Use Fractions?

Fractions, like whole numbers, are a useful way of describing the quantities and consistencies around us. As such, they serve as useful tools!

If we divide a smaller number by a larger number (such as 1 by 2), it is helpful to have a way of expressing that quantity ($\frac{1}{2}$). It is also helpful to record quantities as parts of other quantities. Saying you have read 100 pages of a book does not really give a clear idea how close to the end you are (the book could have 200 pages or 500 pages). But saying you have read $\frac{100}{500}$ gives a much better idea. Likewise, $\frac{1}{4}$ cup or $\frac{1}{4}$ yard gives a clear understanding of a specific portion of a cup or a yard.

Fractions aid in working with money. We use fractions to represent portions of a dollar (we typically write these fractions using decimal notation, which we will get to later).

Fractions even have uses in music! Hum the tune to "Amazing Grace" or some other song. We say "the" in "how sweet the sound" rather quickly, but linger a little longer on the word "sound." When we write music for songs, we need a way of specifying how long to hold each note.

Musicians use different notes to specify different lengths of time. The notes are based on a consistent rhythm called a beat. When you clap your hands to a song, you are clapping the beat of the song. The ⬤ symbol, called a whole note, is most commonly used to specify 4 beats.

A variety of other notes represent a fraction of the whole note. A half note is worth $\frac{1}{2}$ of a whole note, or 2 beats ($\frac{1}{2}$ of 4 is 2), a quarter note is worth $\frac{1}{4}$ of a whole note, or 1 beat, and an eighth note is worth $\frac{1}{8}$ of a whole note, or $\frac{1}{2}$ a beat.

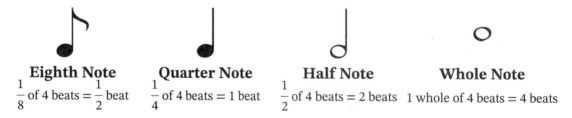

Eighth Note	Quarter Note	Half Note	Whole Note
$\frac{1}{8}$ of 4 beats = $\frac{1}{2}$ beat	$\frac{1}{4}$ of 4 beats = 1 beat	$\frac{1}{2}$ of 4 beats = 2 beats	1 whole of 4 beats = 4 beats

Sometimes songwriters will decide to make the whole note worth 2 beats instead of 4. In this case, a half note is still worth $\frac{1}{2}$ of the whole note. But since the whole note is now worth 2 beats instead of 4, the half note is now worth $\frac{1}{2}$ of 2, or 1. Although many musicians do not realize it, being comfortable with fractions can help them understand and, most especially, write music.

The application of fractions in music is just one example of how we can use math to help us praise the Lord, serve and encourage others, and simply refresh ourselves.

We use fractions because, like whole numbers, they help us describe quantities and consistencies God placed around us.

Revealing Numerators and Denominators — Fractions Throughout History

While today we put the number representing the number of parts (called the *numerator*) on top of a line and the number representing the total parts (called the *denominator*) on the bottom of a line, this was not always the case! Throughout history, men have expressed partial quantities in various ways.

Egyptian — The Egyptians had more than one way of expressing $\frac{1}{4}$. In fact, according to Florian Cajori, "there are three forms of Egyptian numerals: the hieroglyphic, hieratic, and demotic."[55] The symbols for $\frac{1}{4}$ in these three forms are shown below. The first one on the left is the hieroglyphic form, the middle one the hieratic, and the one on the right the demotic.[56]

Babylonian — The Babylonian number system was based on 60. They wrote $\frac{1}{4}$ by writing 15, which is $\frac{1}{4}$ of 60. This could get confusing, however, as this symbol could also mean 15.

Greek — According to Florian Cajori, "Greek writers often express fractional values in words...When expressed in symbols, fractions were often denoted by first writing the numerator marked with an accent, then the denominator marked with two accents and written twice."[57] The picture below shows $\frac{1}{4}$ in Greek symbols.

$$\alpha' \; \delta'' \; \delta''$$

---

55. Cajori, *History of Mathematical Notations*, 1:11.
56. Cajori, *History of Mathematical Notations*, 1:11–15. See also Smith, *History of Mathematics*, 2:45–47.
57. Cajori, *History of Mathematical Notations*, 1:26.

Roman — The Romans avoided fractions by using smaller and smaller units of measure. For example, rather than saying they had $\frac{1}{4}$ of an *as* (a specific Roman coin), they would say they had a *quadrans*. Below are some Roman symbols for a *quadrans*, or $\frac{1}{4}$.[58]

Seeing the different ways other cultures recorded fractions helps guard against thinking of our modern method of math as math itself. It reminds us our current fractional notation is just *one* system we can use to help describe God's complex universe! Terms like *numerator* and *denominator* give us a convenient way to reference different parts of fractions.

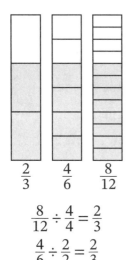

Revealing Equivalent Fractions — Different Ways to Look at Quantities

The same quantity can be represented many different ways. Notice how the shaded area of these rectangles can be looked at as $\frac{2}{3}, \frac{4}{6},$ and $\frac{8}{12}$. All these fractions have the same value — they represent the same portion of a rectangle. We call them equivalent.

We can find equivalent fractions by multiplying or dividing both the numerator and the denominator by the same number. Why? Because when we multiply or divide both the numerator and the denominator by the same quantity, we are essentially multiplying or dividing by 1. Any number divided by itself equals 1 ($4 \div 4 = 1$, $5 \div 5 = 1$, etc.). Because we are multiplying or dividing the fraction by a fraction worth 1, we are not affecting the value of the fraction; we are simply expressing it a different way.

Revealing Reducing Fractions

Expressing quantities using the smallest numbers possible makes them easier to process. Notice how it is easier to quantify $\frac{2}{3}$ than $\frac{8}{12}$, and $\frac{1}{2}$ than $\frac{26}{52}$ or $\frac{58}{116}$. Reducing (or simplifying) fractions is a fancy term for expressing a fractional quantity in the lowest form possible.

Revealing Proper Fractions, Improper Fractions, and Mixed Numbers

Sometimes, we have a numerator that is greater than the denominator. For example, suppose we had 15 bite-size candies and wanted to divide them between 2 people. Since we can use fractions to represent division (in fact, in upper math, the fraction line basically replaces the division symbol), we can write this situation like this: $\frac{15}{2}$. We call this an

---

58. Cajori, *History of Mathematical Notations*, 1:36. See also Smith, *History of Mathematics*, 2:208–209.

improper fraction because we're dividing a greater quantity by a lesser quantity rather than representing a partial quantity (which we call a **proper fraction**).

While improper fractions prove very useful in upper math and science applications, in everyday applications, it often aids in instantly understanding what quantity is being represented if we go ahead and rewrite improper fractions as what we call **mixed numbers**. To rewrite as a mixed number, we go ahead and complete the division, leaving any remainder as a fraction.

For example, we can rewrite $\frac{15}{2}$ as $7\frac{1}{2}$. All we did was complete the part of the division we were able to complete. While the division can be done mentally, we'll show it using the long division algorithm.

$$\begin{array}{r} 7 \\ 2\overline{)15} \\ -14 \\ \hline 1 \end{array}$$

Now, 1 doesn't divide by 2 ... but we can write the answer as a fraction! We still have 1 to divide by 2, which we can show as $\frac{1}{2}$. That fraction line means division...and it also shows the result of that division. If we could take 1 candy and divide it between 2 people, each person would receive $\frac{1}{2}$ of the candy. Notice that $\frac{1}{2}$ times 2 equals 1 ....we don't have anything left to divide.

$$\begin{array}{r} 7\frac{1}{2} \\ 2\overline{)15} \\ -14 \\ \hline 1 \\ -1 \\ \hline 0 \end{array}$$

Note that mixed numbers make the most sense when students realize that the fraction line represents division. When we rewrite an improper fraction as a mixed number, we're just completing part of the division.

Knowing this also makes it easy to convert from a mixed number back to an improper fraction. All we need to do is rewrite the whole number portion as a division by the denominator of the fraction. To rewrite $7\frac{1}{2}$ as an improper fraction, we'd multiply 7 by 2, getting 14. This tells us that we could rewrite 7 as $\frac{14}{2}$. (Stop and do the division; 14 divided by 2 equals 7.) Then we add the additional 1 that we have in $\frac{1}{2}$, giving us $\frac{15}{2}$.

Conclusion

Fractions are yet another way of looking at and exploring the quantities around us. Rather than looking at a piece of pie as a single unit (1 unit), fractions allow us to easily represent what part of a whole that piece represents ($\frac{1}{8}$). And because of the amazingly consistent way God causes objects to divide, we can use fractions to represent division too. Fractions, like whole numbers, help us name and explore God's creation.

TEACHING SUGGESTIONS AND IDEAS

Objective: *To continue to build a biblical understanding of math by presenting fractions as a useful system for recording both partial quantities and division.*

Specific Points to Communicate:

- *Fractions, like whole numbers, are a system for naming quantities.*
- *Our modern fractional notation is not the only system for naming partial quantities or division.*
- *Quantities can be expressed different ways depending on the need.*

You will most likely teach fractions in conjunction with other math concepts, such as division, ratio, decimals, and percents. Please see those respective chapters for further ideas.

Most modern math books present fractions exclusively as ways of representing a partial quantity, without letting students know they also represent division. I would recommend breaking this trend, as an understanding of fraction's relationship to division is crucial for really understanding why we work with fractions the way we do. It also lays the proper groundwork for understanding upper-level concepts. Without this understanding, students are left merely memorizing rules, which eventually leads to a wrong perspective of math itself (a list of rules rather than a tool to describe quantities).

Example

Your textbook, like the one below, will most likely demonstrate fractions with shapes, giving the child plenty of practice identifying shaded portions of different shapes.

> *What fraction of each circle is shaded? Match the circles to the correct answers.*[59]

While identifying portions of circles can be helpful, so much more can be done with fractions! If a child only colors shapes and completes problems on paper every day, he can easily view fractions as busywork, thus missing out on really learning fractions as God-given tools.

To help your child see fractions as a way to name real-life quantities, take him on a discovery process. *Show* him how to use fractions to name and explore the quantities around him. Have him practice naming real-life quantities different ways (a piece of pie can be thought of as 1 piece of pie or $\frac{1}{8}$ of a pie, etc.). Let him play around with different

---

59. Primary Mathematics Project Team, *Primary Mathematics 3B*, Workbook, Part 2, 5.

ways to write partial quantities and results of division so he will see the method taught in his textbook as *one* method out of many. The "Ideas" section lists lots of ways you could integrate your study of fractions into everyday life.

You can also show fraction's usefulness in recording division problems. One way to present the division aspect of fractions would be to take 15 pretzels and have your child divide them into two equal piles. He should end up with 1 pretzel left over. Ask him how to break or divide it equally between the two piles, then have him think about what he just did. He took 1 pretzel and broke it into 2. He has already learned to express that as 1 ÷ 2. But how would we represent the answer to 1 ÷ 2? Guide him into writing $\frac{1}{2}$, pointing out that this method of recording division tells us both what we are dividing (1 ÷ 2) and our answer (a half of something). You could then have him practice writing lots of different division problems as fractions, reminding him that, because of the consistent way God holds all things together, 2 ÷ 4 equals the same quantity as 2 parts out of 4.

You might also want to have your child practice rewriting remainders to division problems as fractions. Notice that in 25 ÷ 3, we end up with 1 we need to divide by 3. Rather than writing "r1" to stand for a remainder of 1, we could write $\frac{1}{3}$, which means the same thing as 1 ÷ 3.

$$\begin{array}{r} 8\ \text{r}1 \\ 3\overline{)25} \\ -24 \\ \hline 01 \end{array}$$

Ideas

- ◆ **Have your child practice viewing quantities as fractions of other quantities.** Since any number can be written as a fraction of another number, fractions have lots of uses!

  » **At the Library** — When you check out books from the library, look at each book as a fraction of the total number of books. (If you checked out 12 books, each one is $\frac{1}{12}$ of the total books.)

  I know you normally would not bother to think about a book from the library as $\frac{1}{12}$ of your total library books, but you are trying to show your child he could view anything as a fraction, thereby training him to think mathematically and view fractions as ways of describing quantities around us.

  You might also examine books of varying lengths and illustrate how merely saying we have finished reading 100 pages of a book does not really describe how close we are to the end of the book. Demonstrate how we could write the number of pages as a portion of the total pages using fractions ($\frac{100}{110}$ if we'd read 100 out of 110 pages, $\frac{100}{300}$ if we'd read 100 out of 300 pages, or whatever the case might be).

  » **In the Kitchen** — When you bake cookies, talk with your child about viewing cookies as a fraction of the total. (If you baked 36 cookies, you could look at 4 cookies as $\frac{4}{36}$ of your total cookies.)

» **During the Day** — Ask your child what fraction of his math worksheet he has finished. (If he has finished 4 out of 20 problems, he could represent that as $\frac{4}{20}$.) Again, the point is to help him understand he can view anything as a fraction, or part, of another quantity.

» **At a Game** — If you go to a football game, notice the 50-yard line divides the football field into halves, each one of which is 1 part out of 2 ($\frac{1}{2}$ of the total field).

◆ **Look for ways to display the usefulness of fractions.** While we will explore the usefulness of fractions in more depth in the next chapter, you can begin pointing it out right from the beginning. Look for ways you use fractions (cooking, sewing, shopping, building, measuring, etc.) and share them with your child. If your grocery store has a manual produce scale, let your child weigh some fruits or vegetables, noticing the fractional markings ($\frac{1}{2}$ a pound, $\frac{1}{3}$ a pound, etc.). Sometimes liquid medicines or laundry detergent are measured using measuring cups with fractions on them.

◆ **Discuss fraction's role in music with your child.** See the discussion earlier in the chapter for a general overview. You may even want to find a basic music theory book to work through. If your child already plays a musical instrument, get some staff paper and have him write his own song. If he needs help, you might suggest he set a scripture verse to music or think of some emotion (suspenseful, calm, reverent, etc.) he could try to communicate.

◆ **Reduce (simplify) everyday fractions.** Reducing (also known as simplifying) fractions (such as $\frac{12}{24}$ into $\frac{1}{2}$) is one aspect of fractions you will want to be sure you cover. Not only does reducing fractions prove invaluable when adding, subtracting, multiplying, and dividing fractions, it also gives us a better perspective of the partial quantity the fraction is representing. For example, knowing we have completed 144 out of the 288 pages in a book ($\frac{144}{288}$) does not really give us a good idea where we are in the book. But if we simplify this fraction, we realize 144 is really half of 288, meaning we have read $\frac{1}{2}$ of the book. The fraction $\frac{1}{2}$ is a simpler way to view our progress than $\frac{144}{288}$.

◆ **Practice dividing the same-sized rectangle into different-size segments.**
You can use this to show your child that the same quantity (one rectangle) can be expressed many different ways, depending on the need.

$\frac{1}{2}$		$\frac{1}{2}$	
$\frac{1}{3}$	$\frac{1}{3}$		$\frac{1}{3}$
$\frac{1}{4}$	$\frac{1}{4}$	$\frac{1}{4}$	$\frac{1}{4}$

(Continuing with rows for $\frac{1}{5}$ through $\frac{1}{9}$, each divided into that many equal segments.)

PARTING NOTE

Fractions require more abstract thinking than the other math concepts we have covered so far. As Ron Aharoni acknowledged, "their depth is indeed on a par with that of many university topics."[60] So please do not be discouraged if your child does not grasp them right away.

Keep running to the Lord and asking Him for ideas. You may find you need to present something in a different way...or even hold off for awhile. Whatever the case, let God use it to teach both you and your child eternal lessons as you work through math.

> *For the LORD giveth wisdom: out of his mouth cometh knowledge and understanding.*
>
> PROVERBS 2:6 (KJV)

---

60. Aharoni, *Arithmetic for Parents*, 124.

Fractions:
Operations

As we have already seen, whole quantities add, subtract, multiply, and divide in a consistent fashion, making it possible to memorize and develop methods to perform these operations on paper. In this chapter, we will explore how fractions also add, subtract, multiply, and divide in a consistent, predictable fashion.

We add and subtract fractions by making sure the denominators are the same and adding or subtracting the numerators.

$$\frac{2}{4} + \frac{1}{4} = \frac{2+1}{4} = \frac{3}{4}$$

We multiply fractions by multiplying both the numerators and denominators.

$$\frac{2}{3} \times \frac{3}{4} = \frac{2 \times 3}{3 \times 4} = \frac{6}{12}$$

We divide fractions by inverting the second fraction and multiplying both the numerators and denominators.

$$\frac{2}{3} \div \frac{3}{4} = \frac{2}{3} \times \frac{4}{3} = \frac{2 \times 4}{3 \times 3} = \frac{8}{9}$$

Sounds pretty factual — but why do these "rules" work? Why do we add only the numerators when adding fractions ... and invert and multiply to divide them? Why do we bother to work with fractions anyway?

To answer these questions, let us review some biblical principles and explore what these "rules" really represent and how they prove useful.

Why Do These Rules Work?

The biblical principles we explored for whole numbers also apply to fractions. We can only reduce the addition, subtraction, multiplication, and division of fractions to "rules" because quantities add, subtract, multiply, and divide in a consistent fashion. If this universe were not consistent, working with fractions would have no use outside a textbook — fractions would be a mere intellectual game.

As we saw when we explored adding whole numbers, the Bible gives us an explanation for why the universe operates so predictably. It tells us God rules all things and never changes. He upholds all things consistently. As we explore the intricacies of fractions, we see to an even deeper level the consistency around us and are, in turn, reminded of God's faithfulness.

Psalm 119:90–91 reminds us God established the earth and the "ordinances" around us. Creation continues to operate according to these ordinances because all things serve God.

Thy faithfulness is unto all generations: thou hast established the earth, and it abideth. They continue this day according to thine ordinances: for all are thy servants.

PSALM 119:90–91 (KJV)

The Tool Behind the Rules

So how exactly do the rules for working with fractions describe a real-life consistency? To answer that question, join me in taking a look at several rules individually. These rules are not arbitrary at all, but rather ways of describing a consistency around us using the notation for fractions we have adopted.

Addition and Subtraction

The "rule" for adding and subtracting fractions is to rewrite the equation so the denominators are equal, then add or subtract the numerators. Why? Let us examine a real-life situation to see.

If you have ever made yeast bread, you have probably added flour to a mixing bowl a little at a time. Say you added a $\frac{1}{2}$ cup, then a $\frac{1}{4}$ cup. How would you figure out the total amount of flour used?

You cannot just add the denominators and numerators. As the picture shows, $\frac{1}{2} + \frac{1}{4}$ does not equal $\frac{2}{6}$. To find the answer to $\frac{1}{2} + \frac{1}{4}$, we have to first rewrite the equation so the denominators are equal.

$$\frac{1}{2} + \frac{1}{4} \neq \frac{2}{6}$$

however

$$\frac{2}{4} + \frac{1}{4} = \frac{3}{4}$$

Note that $\frac{1}{2} = \frac{2}{4}$; we've just rewritten the equation with a common denominator.

We have to rewrite a fraction with the same denominator because we need to first look at them as portions of or divisions by the same quantity. Once expressed this way, then we could add together the portions — that is, the numerators. Then we will be able to express their addition as a portion of or division by the same quantity as well.

Subtraction works much the same way. In order to subtract parts of a whole, we have to subtract equal parts. The "rule" is just an easy way of expressing this. Let us say we went to the sewing store and found a bolt of fabric with $\frac{2}{3}$ yard on it. We know our pattern only calls for $\frac{1}{4}$ yard, but want to have some left over to make various things. Subtracting $\frac{1}{4}$ yard from $\frac{2}{3}$ yard will tell us how much we will have left. To subtract, though, we have to look at both quantities as parts of or divisions by the same quantity. Then we subtract the numerators to see how many parts of or divisions by that quantity we have.

$$\frac{2}{3} - \frac{1}{4} = \frac{8}{12} - \frac{3}{12} = \frac{5}{12}$$

Multiplication

We multiply fractions by multiplying the numerators and the denominators.[61] To better understand this rule, let us explore what we are really representing when we multiply fractions.

Multiplying fractions, like multiplying whole numbers, is a "shortcut" for adding repeated quantities in one step.

$3 \times \frac{1}{2}$ is a shortcut for writing $\frac{1}{2} + \frac{1}{2} + \frac{1}{2}$ ($\frac{1}{2}$ taken 3 times); $2 \times \frac{2}{3}$ is a shortcut for writing $\frac{2}{3} + \frac{2}{3}$ ($\frac{2}{3}$ taken 2 times).

$3 \times \frac{1}{2}$ $2 \times \frac{2}{3}$

Just as with whole numbers, we use the first number in a multiplication problem to stand for the number of times we are taking the second number.

Keeping this in mind, any guesses what $\frac{1}{4} \times 2$ represents? $\frac{1}{4} \times 2$ represents 2 taken $\frac{1}{4}$ times, or, in English, $\frac{1}{4}$ of 2 (what we would get if we divided 2 wholes into 4 parts). Notice how this is the same quantity (shown by the grey area) as $2 \times \frac{1}{4}$ or $\frac{1}{4} + \frac{1}{4}$ — just as with whole numbers, the order in which we write the multiplication problem does not really matter because they both represent the same quantity.

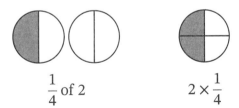

$\frac{1}{4}$ of 2 $2 \times \frac{1}{4}$

Notice I substituted the word "of" for the times sign in the first equation above; the word "of" can help us remember that when we multiply by a fractional quantity, we're finding a portion "of" another quantity.

---

61. Note: If there is a fraction in the denominator, multiplication gets a bit more complicated.

Well, that was easy enough! Now let us look at multiplying two fractions. Consider $\frac{1}{4} \times \frac{1}{2}$. This represents a quantity worth $\frac{1}{4}$ of $\frac{1}{2}$ (or what we would get if we divided half a quantity into 4). For example, it represents what we would get if we were to take a pie and half it, then divide that half into 4. You can see from the picture that $\frac{1}{4}$ of $\frac{1}{2}$ equals $\frac{1}{8}$ (if $\frac{1}{2}$ is in 4 parts, the half would have 4 parts as well, for a total of 8).

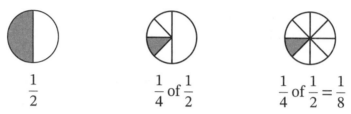

With this grasp of what multiplication represents, let us see if we can better understand how the "rule" of multiplying the numerators and the denominators accurately describes what happens in real life.

Start by multiplying a whole number by a fraction. Remember, multiplying $2 \times \frac{1}{4}$ means the same thing as $\frac{1}{4} + \frac{1}{4}$. Multiplying 2 by the numerator will give us the desired answer — the value of $\frac{1}{4}$ taken 2 times. Since whole numbers can be rewritten as fractions of one ($\frac{2}{1}$ means the same thing as 2, since $\frac{2}{1}$ means $2 \div 1$, and dividing a number by 1 does not change the value), we could have rewritten our equation as $\frac{2}{1} \times \frac{1}{4}$ and followed the "rule" of multiplying the numerators and the denominators.

Now consider multiplying a fraction by a fraction. Take $\frac{3}{6} \times \frac{2}{4}$ for example. Here we want to find $\frac{3}{6}$ of $\frac{2}{4}$, or what we would get if we started with $\frac{2}{4}$, and took $\frac{3}{6}$ of it. How do we take $\frac{3}{6}$? Remember, fractions are a way of writing division. Thus $\frac{3}{6}$ means $3 \div 6$. To take $\frac{3}{6}$, we need to divide $\frac{2}{4}$ by 6, then take 3 of those parts (multiply by 3).

Notice how this equals the same quantity as we would have obtained if we had multiplied the numerators and the denominators (the grey areas are both $\frac{1}{4}$ of the whole).

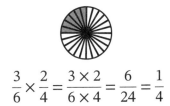

$$\frac{3}{6} \times \frac{2}{4} = \frac{3 \times 2}{6 \times 4} = \frac{6}{24} = \frac{1}{4}$$

A key to understanding multiplying by fractions is to remember what a fraction means. As Ron Aharoni points out, "Since $\frac{4}{5} = 4 \div 5$, multiplying by $\frac{4}{5}$ means multiplying by 4 and dividing by 5.[62]" So if we have $3 \times \frac{4}{5}$, we're multiplying 3 by 4 and dividing that product by 5. If we have $\frac{3}{10} \times \frac{4}{5}$, we're multiplying 3 by 4 and dividing that by the product of 10×5.

Division

To divide fractions, the "rule" is to invert (i.e., use the reciprocal of) the second fraction and multiply. Why? To find out, let us take a few minutes to look at what fractional division represents.

Say we were to divide a pie in half. Since fractions are one way to represent division, we could represent this by writing either $1 \div 2$ or $\frac{1}{2}$.

Now let us say we divide that half a pie into 4 pieces. We have now divided a half by 4. We could write this as $\frac{1}{2} \div 4$. We can see from drawing this out that this equals $\frac{1}{8}$.

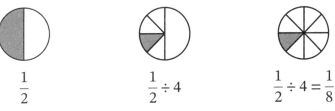

Notice how we are finding the same exact quantity as when we find $\frac{1}{4} \times \frac{1}{2}$ or $\frac{1}{2} \times \frac{1}{4}$.

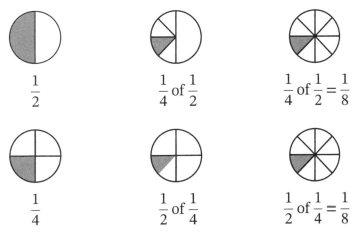

---

62 Ron Aharoni, *Arithmetic for Parents*, 136.

As you can see, $\frac{1}{2} \div 4$ represents the same quantity as $\frac{1}{2}$ of $\frac{1}{4}$, or $\frac{1}{2} \times \frac{1}{4}$.

Keeping this in mind, let us take a look at how we could have found the answer to or original division problem using the "rule." Since dividing a number by 1 does not change the value of a number, we can think of 4 as $4 \div 1$, or $\frac{4}{1}$, and rewrite our equation as $\frac{1}{2} \div \frac{4}{1}$. Following the "rule" to invert the second fraction and multiply would give us $\frac{1}{2} \times \frac{1}{4}$, which we just saw equalled the same thing as $\frac{1}{2} \div 4$. The "invert and multiply" rule is just a shortcut to change an equation we do not know how to solve into an equivalent one we already know how to solve!

Before we move on from division, I would like to briefly mention a different way of looking at division. While we typically think of division as dividing a quantity into smaller parts, we often use it to find out how many smaller parts could fit inside or be contained in a given quantity.

This is easiest to see with whole numbers. When we divide 10 by 2, our answer, 5, tells us both what would happen if we took 10 and divided it into 2 piles, and also how many times 2 is contained inside 10 (5 times). Likewise, when we divide 6 by 2, the answer (3) not only tells us how many we would have if we divide 6 into 2, but also how many times 2 could fit inside 6.

Thinking of dividing fractions this way proves helpful when we come across problems like this: $\frac{1}{2} \div \frac{1}{4}$. While we cannot always physically divide something by a quarter, we can find out how many times $\frac{1}{4}$ fits inside $\frac{1}{2}$ (2 times). Notice there are 2 quarters in a half.

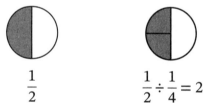

$\frac{1}{2}$ $\frac{1}{2} \div \frac{1}{4} = 2$

Conclusion

Rules about how to add, subtract, multiply, and divide fractions describe consistencies God holds together. They apply practically because of God's unending faithfulness and because He, and He alone, rules over all!

TEACHING SUGGESTIONS AND IDEAS

Objective: *To help the child view multiplication and division of fractions as useful shortcuts that work because of God's faithfulness.*

Specific Points to Communicate:
- *Rules show us how to describe consistencies using a specific notation.*

- *Fractions apply outside a textbook because they describe consistencies God created and sustains.*

The typical math student is bombarded by rules when learning fractions. He is expected to learn rules for simplifying, adding, subtracting, multiplying, and dividing fractions.

But *why* do all these rules work? And *what* are we really doing when we multiply or divide a fraction anyway?

If your child is not directly given the answers to these important questions, he will just be memorizing rules instead of really understanding the principles. As you seek to teach math biblically, one of your goals is to teach your child to use math as a useful tool to serve the Lord. If all he learns is rules, he is not going to be equipped to apply the principle to new situations as they come up. He is also in danger of seeing rules as the mechanisms making math work instead of the tools we use to help our fallen brains record the way God keeps this universe working.

So show the principle behind the rule! Instead of teaching your child to multiply fractions a certain way, show him with household manipulatives or drawings how this rule actually records what is happening in real life. Understanding the principles will help your child view fractions as a way of recording real-life objects and better equip him to use fractions as a useful tool.

As with other concepts, do not be afraid to slow down or work on a different math concept if your child is not getting an aspect of fractions!

Example

Below are two example presentations, both of which present a rote rule to memorize.

> **Addition** —*When you add or subtract fractions with a common denominator, you add or subtract the numerators and keep the common denominator in your answer.*[63]

> **Division** — *Reciprocals are used to divide by a fraction. Dividing by a fraction gives the same answer as multiplying by the fraction's reciprocal. Rewrite division by a fraction as multiplication by its reciprocal.*[64]

Notice neither of these presentations really explain *why* the rules work or *what* we are really doing when we divide or add fractions. You could fix this by taking the time to demonstrate each rule concretely, making sure your child learns the rule with understanding. You do not want him learning to blindly trust a math rule, but rather developing his God-given ability to learn about the consistencies around him.

When teaching addition, rather than jumping right to the rule, you could start from a real-life problem. You could make a loaf of bread together (or some other recipe), having your child use a $\frac{1}{2}$ cup measurer to add the flour, writing down each amount as he adds

---

63. Rasmussen, *Key to Fractions: Adding and Subtracting*, 9.
64. Rasmussen, *Key to Fractions: Multiplying and Dividing*, 31.

it ($\frac{1}{2} + \frac{1}{2} + ...$). Have him figure out how much flour he has added. This should be fairly easy with $\frac{1}{2}$ cups, as it is easy to see that $\frac{1}{2} + \frac{1}{2}$ equals one. You can point out how adding the numerators would get this answer. Then have your child use a $\frac{1}{2}$ cup and a $\frac{1}{4}$ cup measurer and try to figure out how much flour their addition yields. Point out that he could not just add the numerators now, and guide him through how to add fractions with unequal denominators, letting him think of the "rule" himself.

Understanding the principles and seeing them in action can help a child view fractions as a way of recording real-life objects and better equip him to apply fractions as a real-life tool. If you are not quite sure what we are really doing when we add, subtract, multiply, or divide, please see the overview we looked at earlier in the chapter.

One idea to introduce division would be to make a pie and cut it in half. Then cut that half into 4 pieces and ask your child how much of the whole pie each piece represents. Walk through how you started with $\frac{1}{2}$, then divided that half by 4, which we could represent as $\frac{1}{2} \div 4$. Show your child how he could have found the answer on paper by multiplying $\frac{1}{2}$ by the inverse, or reciprocal, of the second number ($\frac{1}{2} \times \frac{1}{4}$), and how we can think of it as multiplying the $\frac{1}{2}$ by 1 (i.e., taking 1 of it) and then dividing it by 4.

You could then go over what division really expresses (how many times one quantity can be broken into another, or how many times one quantity can fit inside another). You could use whole numbers first to demonstrate, then show your child how division of fractions is really the same thing. After the child has grasped the concept, present the rule as a way of describing this consistency.

Ideas

- ◆ **Make cards with your child.** Wouldn't it be fun to make cards to bless others while learning math? Give it a try! Making cards can involve lots of fractions. To start with, the paper needs cut to the right size. If you want to make two cards out of one sheet, you will first need to cut the paper in half. Since a sheet of paper is 11 inches wide, you will need to cut the paper at the $5\frac{1}{2}$ inch mark to cut it in half. You just used a fraction!

  Fractions play a lot of other roles in card making. Say you stamp an image that is about 3 inches by 4 inches. You cut the image out and want to give it a small border — say $\frac{1}{8}$ inch — of a darker color. Using

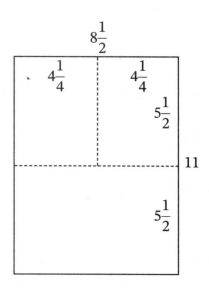

110 REVEALING ARITHMETIC

fractions, you can mentally figure out the size you need to cut the darker colored paper.

Add $\frac{1}{8}$ inch to each side of the image.

$$3 + \frac{1}{8} + \frac{1}{8} = 3\frac{2}{8} = 3\frac{1}{4}$$

$$4 + \frac{1}{8} + \frac{1}{8} = 4\frac{2}{8} = 4\frac{1}{4}$$

The backing needs to be $3\frac{1}{4}$ inches by $4\frac{1}{4}$ inches if you want an $\frac{1}{8}$ inch border.

Or say you want to put a strip of colored paper across the front of the card. You can figure out how wide to make the strip using fractions and division. You know the paper started out $8\frac{1}{2}$ by 11 inches. You cut it in half on the 11 inch side, making it $8\frac{1}{2}$ by $5\frac{1}{2}$ inches ($11 \div 2 = 5\frac{1}{2}$). Then you folded it in half again, making it $4\frac{1}{4}$, since $8\frac{1}{2} \div 2 = 4\frac{1}{4}$, by $5\frac{1}{2}$ inches. You have now figured out the size of your current paper and hence the size strip you need.

- **Explore applications of fractions in measurement.** Fractions are used a lot in measuring (pull out a ruler — inches are parts, or fractions, of a foot, and inches themselves are usually broken down into 16 fractional parts or even 32 parts on some rulers). Consider having your child measure the lengths and widths of various rooms, subtracting to find out how much bigger one length is from another. Or he could measure his height and compare it with someone else's height using subtraction.

- **Have your child help you double or half a recipe, or combine measuring cups to measure ingredients (such as use a $\frac{3}{4}$ cup twice to measure $1\frac{1}{2}$ cups).**
When you double ingredients, you are really multiplying fractions! Doubling $\frac{1}{4}$ cup sugar could be represented as $2 \times \frac{1}{4}$. Halving $\frac{1}{2}$ a cup of sugar could be represented as $\frac{1}{2} \div 2$ or as $\frac{1}{2} \times \frac{1}{2}$. When you use your $\frac{3}{4}$ cup measurer to measure $1\frac{1}{2}$ cups of flour, you are adding fractions ($\frac{3}{4} + \frac{3}{4} = \frac{6}{4} = 1\frac{2}{4} = 1\frac{1}{2}$). You might challenge your child to use as few measuring cups as possible.

- **Have your child apply fractions to his own life.** Finding everyday applications of fractions is not difficult. Here are a few ideas to get you started.
  - » **Grocery Shopping** — Manual grocery scales give the weight in fractions of a pound. Explain to your child the need to know approximately how much your

groceries cost before getting to the checkout. Rounding the cost per pound to the nearest dollar,[65] have your child calculate how much different fruits or vegetables you need would cost. This will involve multiplying fractions ($\frac{1}{2}$ of a pound × $2 per pound = $1). Ask your child to find the cost of $\frac{1}{2}$ and $\frac{1}{4}$ of a pound first, then challenge him a bit more by asking for the cost of larger quantities, such as $\frac{3}{4}$ of a pound.

» **Sewing or Woodworking Projects** — Have your child start a sewing or woodworking project and figure out the yardage or lumber needed. This often involves fractions.

» **Home Improvement Projects** — Fractions often have applications in home improvement projects. For example, the instructions on mixing a solution might tell you to put in $\frac{1}{4}$ cup of a concentrate per quart of water. If you want to make 5 quarts, you could multiply $5 \times \frac{1}{4}$ to find out how much to put in.

» **Car Trips** — Next time you head out on a long car trip, show your child how to use fractions to answer the how-much-farther-do-we-have-to-go question himself! Have him compute how far $\frac{1}{4}$, $\frac{1}{2}$, and $\frac{3}{4}$ of the trip would be. For example, if traveling 600 miles, $\frac{1}{4}$ would be $\frac{1}{4} \times 600$ (or $600 \div 4$), or 150 miles. If you know how far $\frac{1}{4}$ of the trip is, you can easily gage how far along you are. If you have gone 200 miles, you know you are more than $\frac{1}{4}$ of the way there (you're actually $\frac{1}{3}$ of the way there).

» **Nutritional Content of Food** — Sometimes it is very important to know how much sugar, salt, etc., we are eating. Say you make a batch of 8 giant muffins and put $\frac{1}{2}$ cup of sugar into it. You might ask your child to figure out how much sugar he would be eating if he ate 2 muffins. Since there are 8 muffins, 2 muffins are $\frac{2}{8}$ of the total. So to find the sugar, we need to find $\frac{2}{8}$ of $\frac{1}{2}$ cup ($\frac{2}{8} \times \frac{1}{2} = \frac{2}{16} = \frac{1}{8}$). We would be eating $\frac{1}{8}$ a cup of sugar. (Wow! We might decide to eat only one muffin.)

---

65. If you have not yet taught rounding, take a trip to the grocery store to teach it first. Explain how important it is to know approximately how much money you are spending so you do not get surprised at the checkout. Have your child try to keep track of all the groceries you put in the cart without a calculator. He will quickly discover he needs to round! Then explain the mechanics of rounding. When rounding, look at the digit to the right of the place you want to round the number to. If that digit is 5 or greater, you round up; if it is less than 5, you round down. For example, if rounding $5.59 to the nearest dollar, you see that the first decimal place is a 5 so round up to $6. If rounding $5.49, you'd round down to $5 as the first decimal place is a 4. You might also mention that we use the ≈ to show that something approximately equals a value — that is, that we've found a rounded amount.

- **Look at some of the practical problems given in older math books.** See the Book Extras on ChristianPerspective.net for some math books from the 1800s you can access for free online. Many of these books contain practical problems on all aspects of fractions. You could give your child a few problems as reinforcement exercises, or look at them for additional ideas.

- **Study the life of a mathematician, such as Johannes Kepler or Sir Isaac Newton.** See Appendix A for some general questions you could ask as you study. Both of these men, while they had some theological issues, clearly saw math as a way of describing God's creation and as a testimony to the Creator. Many books written about them seek to twist their philosophy and present them as men who helped bring in a naturalistic explanation of the universe. When you find these twists, take advantage of the opportunity to present the truth and teach discernment. While these men did help bring in modern science, they saw nothing naturalistic about it! Whole pages of their "math" books discourse on God's greatness! For example, Newton wrote, "This most beautiful system of the sun, planets, and comets, could only proceed from the counsel and dominion of an intelligent and powerful Being."[66]

- **Teach simplifying fractions while multiplying as a helpful "shortcut."** When multiplying fractions, it often helps to divide the numerator in one fraction and the denominator in the other by the same number (in the problem shown, 5 and 15 were both divided by 5, and 14 and 2 were both divided by 2). Note that this is doing the same thing as dividing the fraction by fractions worth 1 (in this case, $\frac{5}{5}$ and $\frac{2}{2}$). But instead of doing all the multiplication and then dividing, we divide as we go.

$$\frac{\overset{1}{\cancel{5}}}{\underset{7}{\cancel{14}}} \times \frac{\overset{1}{\cancel{2}}}{\underset{3}{\cancel{15}}}$$

Simplifying the fractions before we multiply them together not only makes it easier to multiply fractions, but also saves us from having to do as much simplification on the answer. It is an example of how men used their God-given observational skills and ability to reason to find a time-saving method.

- **Take a look at applications of fractions in science.** Science is filled with applications of fractions. Many scientific formulas contain fractions. Have your child keep his eyes open as he studies science to see what he can find!

---

66. Newton, Sir Isaac, *Principia*, Book III, in *On the Shoulders of Giants*, ed. Hawking, 1157.

PARTING NOTE

There is an old saying that goes, "I cannot see the forest through the trees." In fractions, it is easy to lose sight of the big picture (the forest) amidst the "trees" of rules and problems.

While teaching your child how to work with fractions, do not lose sight of what fractions are in the first place.

Fractions are simply tools used to record and work with division problems and partial quantities. We need to learn how to work with fractions to better apply them. The shortcuts and quick methods employed in the process are just quicker ways to find the answers we need.

Fractions also remind us God holds everything — even partial quantities — together consistently. Fractions we manipulate on paper only correspond with real life because God holds all things together consistently.

Fractions:
Factoring, GCF, and LCM/LCD

When we discussed adding and subtracting fractions in the last chapter, we saw that in order to add and subtract we had to have the same denominator — that is, we have to be looking at the fractions as portions of or divisions by the same quantity.

While it's easy enough to see that we can rewrite $\frac{1}{2}$ as $\frac{2}{4}$ by just multiplying $\frac{1}{2}$ by $\frac{2}{2}$, a quantity worth of 1, it's not always quite so obvious. It's important to learn to look at what we call the **factors** of a quantity. Not only can they help us find least common denominators (i.e., the lowest value we could rewrite multiple denominators as), but factors also help us think through what numbers a value can be evenly divided by, work with fractions in other ways, and prepare for upper math where factoring helps us solve problems we couldn't solve otherwise.

Viewing a Value as a Product of Factors (i.e., as a Multiple)

When they learned to multiply, students likely learned that we call values they multiplied *factors*, and the result the *product*. In $4 \times 5 = 20$, 4 and 5 are the factors, and 20 the product. Factoring is simply viewing a quantity as a product and thinking through what different values can be multiplied together to equal it. Typically, students will be looking for whole-number factors.

For example, if I take 20 candies, I can arrange those candies in 4 piles, getting 5 in each pile. Or I can arrange them in 5 piles, getting 4 in each pile. Or I can arrange them in 2 piles, having 10 in each pile. Or 10 piles, with 2 in each pile. Or — and this one is rather obvious — I could put them in 1 pile, having 20 in that pile. And I could even put them in 20 piles, with 1 in each pile! In other words, I can think of 20 as the result of 4×5, 5×4, 2×10, 10×2, 20×1, or 1×20.

But if I were to try to arrange those 20 candies into 3, 6, 7, 8, 9, 11, 12, 13, 14, 15, 16, 17, 18, or 19 piles, I would not be able to do so evenly, without having any leftovers.

We would say that 1, 2, 4, 5, 10, and 20 are the whole-number **factors** of 20. They're the whole numbers (a whole number is what we call the numbers 1, 2, 3, and so forth) that could be multiplied by another whole number and get 20. Another way of thinking about that is that we could divide 20 by any one of its factors, and it would divide evenly. We could also say that 20 is a **multiple** of 1, 2, 4, 5, 10, and 20 — that is, those values can be multiplied by another value to equal 20.

Being familiar with the factors of a number helps us think through what it can be evenly divided by. This can aid in mental arithmetic. When dividing mentally, you have to think through factors of the number. For example, if you had 20 candies and 9 people, you'd know instantly that you can't evenly divide all the candies.

Factor Trees, the Greatest Common Factor (GCF), and Simplifying Fractions

Techniques like factor trees are simply aids in helping us think through the various factors of a number. After all, we don't want to have to actually try to divide candies all the time! Because of the consistent way God governs all things, we can memorize multiplication and division facts, and use those to help us think through factors on paper.

In a factor tree, we break numbers down into what we call **prime factors**. A **prime number** is a whole number greater than 1 that cannot be evenly divided by any whole number except by itself and 1. Prime factors, not surprisingly, are factors that are prime numbers.

Prime Numbers Under 100
2, 3, 5, 7, 11, 13, 17, 19, 23, 29, 31, 37, 41, 43, 47, 53, 59, 61, 67, 71, 73, 79, 83, 89, 97

If we were to factor 20 using a factor tree, we'd start by thinking of any 2 whole numbers we could multiply together to get 20. (We'll ignore 20 × 1, as that won't help us, since we'd still need to find factors of 20.) Then we'd think of the numbers we could multiply together to get those numbers. And so forth until all we had at the bottom of the tree were prime numbers. Notice how no matter which factors we start with, we arrive at the same prime numbers. We can rewrite 20 as 2 × 2 × 5.

$$
\begin{array}{cc}
20 & 20 \\
\wedge & \wedge \\
2 \times 10 & 4 \times 5 \\
\wedge & \wedge \\
2 \times 5 & 2 \times 2
\end{array}
$$

Knowing the prime factors proves very useful. It lets us know what factors 2 or more numbers share without us having to worry about finding all of the whole number factors. For example, if we rewrite 48 and 20 as a product of their prime factors, we see that 2 × 2 is a part of the multiplication for both numbers; this means that both of them can be evenly divided by 2 × 2, or 4. While we might not have instantly thought of 4 as a factor of 48 since it's not a prime factor, we can easily see non-prime factors by factoring down to the prime factors and then multiplying them together.

$$
\begin{array}{cc}
48 & 20 \\
\wedge & \wedge \\
8 \times 6 & 2 \times 10 \\
\wedge \wedge & \wedge \\
2\times 4 \;\; 2\times 3 & 2 \times 5 \\
\wedge & \\
2 \times 2 & \\
\end{array}
$$

$$48 = \boxed{(2 \times 2)} \times 2 \times 2 \times 3 \qquad 20 = \boxed{(2 \times 2)} \times 5$$

We would call 2 a **common factor** of 20 and 48, and 2 × 2, or 4, the **greatest common factor** of 20 and 48. We don't have to try to find every factor; we can see from looking at the prime factors that it's the greatest factor they both share. We found the greatest common factor by rewriting each numbers as a product of the prime factors and then multiplying any shared prime factors together. We only multiply 2 by itself once because 2 × 2, or 4, is a factor of both numbers, but 2 × 2 × 2, or 8, is only a factor of 48.

The factor tree is an example of men, using their God-given abilities, coming up with techniques to help make a job simpler. When we go to simplify fractions, if we divide both the numerator and the denominator by their greatest common factor, we'll have simplified the fraction as much as possible.

For example, if we had $\frac{20}{48}$, we know we can simplify it by dividing by $\frac{4}{4}$, a fraction worth 1:

$$\frac{20}{48} \div \frac{4}{4} = \frac{5}{12}$$

Notice that $\frac{5}{12}$ is a lot easier to instantly take in than $\frac{20}{48}$ is. We knew to divide by $\frac{4}{4}$ since we saw that 4 was the greatest common factor of 20 (the numerator) and 48 (the denominator). Simplifying helps us more easily communicate about and work with quantities. And notice that it's only possible because of how consistently God governs all things — it's His faithfulness that allows us to rely on 5 divided by 12 equaling the same thing as 20 divided by 48 ... which we were able to find by dividing by a fraction worth 1 (knowing that wouldn't change the value because dividing by 1 doesn't) and by factoring (which required relying on multiplication to work the same way). If multiplication and division weren't consistent, factoring would be meaningless!

Note that viewing a value as a product of its factors also helps us simplify *while* multiplying fractions. For example, let's say we had this multiplication:

$$\frac{20}{23} \times \frac{3}{48}$$

Rather than completing the multiplication and then simplifying to multiply fractions, we can simplify as we go. Let's rewrite 20 and 48 as the product of their prime factors.

$$\frac{2 \times 2 \times 5}{23} \times \frac{3}{2 \times 2 \times 2 \times 2 \times 3}$$

Now since we multiply the numerators together and the denominators together to multiply fractions, we can rewrite this like this:

$$\frac{2 \times 2 \times 3 \times 5}{2 \times 2 \times 2 \times 2 \times 3 \times 23}$$

Notice we arranged the multiplications in ascending order — because of the consistent way God governs all things, order doesn't matter in multiplication. Arranging like this makes it easier to spot common factors.

Now notice that the numerator and the denominator have 2 × 2 × 3. That means that they're both divisible by 2 × 2 × 3, or 12. But rather than completing the multiplication and then dividing by $\frac{12}{12}$, a fraction worth 1, we can just cancel out those common factors now and then multiply. That way, we'll save ourselves some work. Canceling out the common factors is essentially dividing both the numerator *and* the denominator by that same amount — that is, by dividing by a fraction worth 1, which doesn't change the value.

Multiplying and Simplifying at the End	Simplifying While Multiplying
Multiplying first: $$\frac{2 \times 2 \times 3 \times 5}{2 \times 2 \times 2 \times 2 \times 3 \times 23} = \frac{60}{1{,}104}$$ or $$\frac{20}{23} \times \frac{3}{48} = \frac{60}{1{,}104}$$ Simplifying: $$\frac{60}{1{,}104} \div \frac{12}{12} = \frac{5}{92}$$	$$\frac{\cancel{2} \times \cancel{2} \times \cancel{3} \times 5}{\cancel{2} \times \cancel{2} \times 2 \times 2 \times \cancel{3} \times 23} = \frac{5}{92}$$

Note that when simplifying while multiplying, we can often do the simplification in our head rather than writing out the prime factors. Both columns below show the multiplication of $\frac{1}{6} \times \frac{3}{5}$, but in the one on the right, the cross outs show that we've mentally thought through that 3 is a factor of 6 and so that part of the division will be canceled out by the multiplication by 3 in the numerator. We write a 2 where the 6 was as we know that 6 ÷ 3 is 2, so we'd still have a 2 to divide (we could have rewritten 6 as 2 × 3). In the numerator, we write a 1 instead of the 3 since that multiplication by 3 has canceled out with the 3 that was a factor of 6 in the denominator. Note that we could rewrite 3 as 3 × 1 as we're really thinking of 3 as 3 × 1 (any number times 1 equals itself); if we were to cancel out the 3, we'd be left with 1.

Writing Out the Common Factors	Performing the Division of Common Factors Mentally
$$\frac{1}{6} \times \frac{3}{5} = \frac{1}{2 \times \cancel{3}} \times \frac{\cancel{3}}{5} = \frac{1}{10}$$	$$\frac{1}{\cancel{6}_2} \times \frac{\cancel{3}^1}{5} = \frac{1}{10}$$

The point is that viewing a number as the product of factors helps us simplify fractions, and when multiplying, that simplification can be done as we go. No matter how we do it, when we cancel out common factors from the numerator and the denominator, we're really dividing both the numerator and the denominator by the same amount, thereby dividing by a fraction worth 1 and forming an equivalent fraction. We can rely on it to be an equivalent fraction because of the consistent way God governs all things.

The Least or Lowest Common Multiple or Denominator (LCM/LCD) – More with Adding and Subtracting Fractions

Because of the consistent way God governs all things, the time it takes you to travel a certain distance equals that distance divided by the speed at which you move. So we can write the time as a fraction of distance over speed (remember, fractions can describe division). Suppose you travel a distance of 3 miles at a speed of 5 miles per hour, a distance of 4 miles at a speed of 7 miles per hour, and a distance of 1 mile at a speed of 2 miles per hour. Time it takes to travel each distance in hours could then be written as $\frac{3}{5}, \frac{4}{7},$ and $\frac{1}{2}$.

If we want to add up all of these values to find the total time to travel all three distances together, we would need to first rewrite them all with a common denominator. But what common denominator can we use? We'll save ourselves some time if we use the **least common multiple** of the denominators—that is, the smallest number that all the denominators are factors of. Note that this is also called the **lowest common multiple**... and some books, when focusing on fractions and dealing specifically with denominators, will call it the **lowest** or **least common denominator**. While the consistencies of God's creation don't change, the words we use to describe it can.

In this case, the denominator to use is easy to spot: 5, 7, and 2 are all prime numbers, so the least common multiple will be the result of multiplying them together.

$$5 \times 7 \times 2 = 70$$

These fractions can all be rewritten with a denominator of 70.

In other cases, though, we'd want to first rewrite the denominators as a product of their prime factors in order to find the least common multiple. For example, let's say the times had been $\frac{3}{5}, \frac{7}{20},$ and $\frac{3}{16}$. We'd be dealing with really large values if we rewrote them as equivalent fractions with $5 \times 20 \times 16$, or 1,600, as the denominator! Instead, let's rewrite each denominator as a product of its prime factors.

$$\frac{3}{5} + \frac{7}{2 \times 2 \times 5} + \frac{3}{2 \times 2 \times 2 \times 2}$$

Now, let's find the least common denominator. The first denominator is 5, so we know we need a value that is a multiple of 5. The next denominator can be written as $2 \times 2 \times 5$. So we know we need a value that is a multiple of $2 \times 2 \times 5$. But note that 5 is a factor of $2 \times 2 \times 5$ and the 5 in the first denominator. Whatever value is a multiple of $2 \times 2 \times 5$ will also be a multiple of 5, so we don't have to multiply by 5 twice—just once. Now let's look at the last fraction. There we have $2 \times 2 \times 2 \times 2$ in the denominator. So we need a denominator that is a multiple of $2 \times 2 \times 2 \times 2$. But we don't have to multiply $2 \times 2 \times 5$ by $2 \times 2 \times 2 \times 2$, as $2 \times 2 \times 5$ already has 2×2 in it. We only need to multiply by an additional 2×2, giving us $2 \times 2 \times 2 \times 2 \times 5$, or 80...which is a lot easier to work with than 1,600.

In other words, **we only need to include from each factor the greatest number of times they are in one fraction**. Doing so will make sure that we find the value that is the least common multiple of them all. Then, to solve the problem, we only need to multiply each fraction by a fraction worth 1 that has the remaining factors that denominator needs to get to 80.

$$\frac{3}{5}\left(\frac{2 \times 2 \times 2 \times 2}{2 \times 2 \times 2 \times 2}\right) + \frac{7}{2 \times 2 \times 5}\left(\frac{2 \times 2}{2 \times 2}\right) + \frac{3}{2 \times 2 \times 2 \times 2}\left(\frac{5}{5}\right) = \frac{48}{80} + \frac{28}{80} + \frac{15}{80}$$

TEACHING SUGGESTIONS AND IDEAS

Objective: *To help students see that they can look at numbers as the product of other numbers and use this skill to solve real-life problems.*

Specific Points to Communicate:

- *We can look at quantities as the products of multiplication.*
- *Factoring and the related concepts are only possible because of the consistent way God governs all things.*
- *Factoring helps us think through how numbers can be evenly divided. It aids us in both simplifying fractions and rewriting fractions with common denominators. And it has extensive applications in algebra, aiding us in solving real-life problems there.*

A key to revealing these concepts for students is to make sure they're seeing them outside of a textbook too. Show them how they can think of quantities as a product of other quantities, as we did using candies earlier in this chapter. Give them real-life examples of what the fractions they're simplifying, adding, and subtracting could be representing (see the "Ideas" section to get you started).

And as you teach them the mechanics, keep bringing them back to why it all works: because God is faithfully governing all things. Factoring would be pointless if multiplication and division didn't always work the same way...and they're only reliable because of how faithfully God governs all things!

You might also point out to students that God created a complex universe. And as they move on in math, they're going to explore it in more depth. As they do so, they'll find themselves using fractions a lot. They'll also discover that being able to think of a value as the product of factors proves really useful in helping us solve many problems that don't even have fractions in them. While factoring may seem hard at first, it's helping to train them to dig deeper in their explorations of the quantities God has created.

Example

Most textbooks present the facts about factoring, factor trees, etc., and then have students practice the skills on a variety of problems, including ones involving simplifying and adding/subtracting fractions. It's very easy for students to get lost and view it all as busywork. What do they really care what the prime factors are? Or why bother to add $\frac{1}{5} + \frac{6}{25} + \frac{3}{8}$?

When presenting factoring, you might start with a real-life scenario involving 2 fractions and show your child how factoring can help in comparing them. For example, ask them to imagine they have one investment option that earns $7 every 12 months and another that earns $20 every 36 months. Which is better? They can figure out the monthly amount by dividing the total by the number of months, giving them $\frac{\$7}{12}$ and $\frac{\$20}{36}$. Explain that rather than completing this division, if they had the same denominator, it'd be easy to compare. We want what we call a common denominator. Then explain that an easy way to find common denominators is to think through what we call the factors of the denominators. Have them think through different multiplications that equal 36 until they see they need to multiply $\frac{\$7}{12}$ by $\frac{3}{3}$ to get a denominator of 36 and compare the fractions. ($\frac{\$7}{12} \times \frac{3}{3} = \frac{\$21}{36}$).

You can then explain that they're going to learn a technique to help find common denominators called factoring. Go on to explain how we can think of a number as a product of other numbers. In fact, grab 36 manipulatives and have them think of them as 3 piles of 12 (i.e., 3 × 12), 6 piles of 6 (i.e., 6 × 6), and so forth. Show them how rather than forming piles with manipulatives, we can use factor trees to help us find the factors on paper. Point out that we can only factor on paper because of how faithfully God governs all things — we don't have to physically regroup quantities to see how they will divide because objects divide consistently.

Note that you've now presented factoring's usefulness and connected it to real life, as well as pointed your child to God's faithfulness.

Ideas

- **Demonstrate using actual manipulatives how quantities can be thought of as products of factors.** For example, grab 20 dry beans (or use small candies for a sweet treat!) and have your child count them. Mention that we have 1 set of 20, which we could write as 1 × 20. Then have them divide them into 5 piles. Have them count the number in each pile (4). Point out that 4 × 5 equals 20 (a fact we can rely on because of the consistent way God governs all things). Now mix up all the beans again. Have them divide the beans into 4 piles, counting the number in each pile (5). Explain that we can call 4 and 5 factors of 20, since when we multiply 4 and 5 together, we get 20. Now have them divide the beans into 2 piles. There are now 10 in each pile. 10 and 2 are also factors of 20, as we can multiply them together to get 20. On the flip side, we could call 20 a multiple of 1, 2, 4, 5, 10, and 20, since they can all be multiplied by another number to equal 20. The point of the exercise is to help them see that quantities can be thought of as the result of multiplication—and we can think through the different values we could multiply together to get that quantity. You might also have them try other values to see that they are not factors of 20; we can't divide 20 evenly into 3, for example. You might also explore ways where knowing what a number can be divided by proves useful—see the Book Extras page on ChristianPerspective.net for a link to a resource with some ideas.

- **Explore hours, minutes, and the Babylonian number system.** To understand why having 60 minutes in an hour proves useful, have your child factor 60 and try to list all the factors of it (they should find the prime factors and then multiply different ones of them together to find other factors). Point out how many different factors there are: 1, 2, 3, 4, 5, 6, 10, 12, 15, 20, 30, 60. That means that 60 is divisible by a lot of different numbers! In fact, "it has more divisors than any smaller positive integer.[67]" For comparison, have your child factor 100. Its factors are 1, 2, 4, 5, 10, 20,

---

[67] Roni Jacobson, "60: Behind Every Second, Millenniums of History" (New York Times, 2013), https://www.nytimes.com/2013/07/09/science/60-behind-every-second-millenniums-of-history.html

25, 50, and 100. Even though that's still a lot of factors, it's fewer than the number of factors of 60. Knowing those factors help us simplify. For example, we refer to 30 minutes as half an hour. After all, $\frac{30}{60}$ simplifies to $\frac{1}{2}$. Point out how hard it would be if there were 57 minutes in an hour instead. A half hour would be 28.5 minutes instead of an even 30. You could also look at the history of keeping track of hours with your child, reminding them that God created time and gave us heavenly bodies that mark its passage (see Genesis 1:14-18). Throughout history, men have strove to find systems that tie with the passage of time and movement of heavenly bodies that God ordained. When looking at the history of time keeping, you'll read about the Babylonians; they used 60 as the base for their entire number system, so you could also explore their system (see Appendix B). Point out how before the days of calculators (and even before paper was readily available) when more math needed performed mentally, having values that could be easily divided by other values (and easily simplified) was especially important.

- ◆ **Explore breaking up a circle into degrees.**
  In geometry, students will learn that we can measure a circle by thinking of it as being 360°. Pull out a protractor and show your child this. Explain that as they continue in math, they'll use degrees a lot. For example, we can describe the angle at which we want the roof to tilt in degrees.

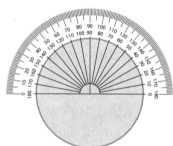

  Understanding factoring helps us understand why 360 makes sense, since it has a lot of factors. Have your child factor 360. Like 60, it has a lot of factors including the same ones for 60 which are 1, 2, 3, 4, 5, 6, 10, 12, 15, 20, 30, 60 (360 equals 60 times 6), as well as 8, 9, 18, 24, 36, 40, 45, 72, 90, 120, 180, and 360. To illustrate that this is a lot of factors, have the student factor 100 (see previous idea) and 75 (factors are 1, 3, 5, 15, 25, and 75). Point out how 360 has more factors—it's divisible by more quantities.

- ◆ **Have students apply factoring to actual examples of simplifying fractions.**
  It's easy to get lost factoring without realizing the purpose—and finding the least common denominator/greatest common factor in order to work with fractions. Give students actual examples of fractions they might need to work with. To do this, think of the fraction line as division. Then you can turn basically any division problem into a simplification problem! For example, the length of a rectangle equals its area divided by its width. If its area is 122 square feet and its width is 6 feet, you'd have this $\frac{122}{6}$. (Note that we're purposefully leaving off units of measure to not complicate things for students just learning the skill.) Ask your child to simplify this so it's easier to read. They could do this by rewriting the numerator and the denominator as the product of prime factors $\frac{61 \times 2}{2 \times 3}$. Now it's obvious that 2 is the greatest common factor, and that this simplifies down to $\frac{61}{3}$. One simple

way to have your student apply factoring is by simply adding example meanings to whatever problems your textbook has given. For example, if your textbook has a problem of simplifying $\frac{85}{15}$, tell them it could be representing the area of a rectangle divided by the width, and the answer will tell you the length. Below are several different examples of real-life relationships involving division that you can use as example meanings for problems in your textbook. Be sure to point out that these relationships hold true because of the consistent way God governs all things! Don't worry if students don't know what some of these things are; you can explain that they'll learn more in future science or math courses.

$$\text{length of a rectangle} = \frac{\text{Area of the rectangle}}{\text{width of the rectangle}}$$

$$\text{width of a rectangle} = \frac{\text{Area of the rectangle}}{\text{length of the rectangle}}$$

$$\text{hourly wage} = \frac{\text{total money made in a week}}{\text{number of hours worked}}$$

$$\text{hours worked in a week} = \frac{\text{total money made in a week}}{\text{hourly wage}}$$

$$\text{monthly price of a subscription} = \frac{\text{yearly price of a subscription}}{12}$$

$$\text{monthly wage} = \frac{\text{total money made in a year}}{12}$$

$$\text{average speed} = \frac{\text{distance traveled}}{\text{time traveled}}$$

$$\text{time traveled} = \frac{\text{distance traveled}}{\text{average speed}}$$

$$\text{acceleration of a ball} = \frac{\text{net force on ball}}{\text{mass of a ball}}$$

$$\text{mass of a ball} = \frac{\text{net force on ball}}{\text{acceleration of a ball}}$$

$$\text{voltage} = \frac{\text{power}}{\text{current}}$$

$$\text{current} = \frac{\text{power}}{\text{voltage}}$$

◆ **For additional ideas, just think of things that you divide and remember that the fraction line represents division.** Have students apply factoring to actual examples of adding and subtracting fractions using the problems your textbook gives, only offer example meanings. Look at the list of real-life relationships given in the preceding idea. If the student has to add 2 fractions together, suggest they could be adding 2 different speeds together...or 2 different monthly wages, hours worked 2 different weeks, etc.

Let students know what they're learning is laying the foundation for more applications down the road. Factoring is a skill that applies down the road in solving real-life problems. Unfortunately, it's one that also requires other skills students probably don't have at this level. But be sure to let them know that factoring has applications even beyond simplifying and adding/subtracting fractions.

PARTING NOTE

As students start working with factoring and fractions, it's easy to get bogged down and forget the purpose of why they're learning various skills. You'll want to be sure to incorporate what applications you can, and to remind them again that it's God who keeps multiplication (which factoring is based on) in place.

Decimals

Most of us are exposed to decimals from the time we are little, mainly because we use them to write parts of a dollar. But what exactly are decimals? Why do they work? What is their purpose? Join me in taking a look!

Decimals — A Shorter Notation

Decimals are a different notation we use to describe on paper the quantities and consistencies God placed around us. Like fractions, they are a way of "naming" partial quantities, helping us express very tiny quantities (like the size of an electron) and record very precise results of a division problem.

To better understand how decimals describe partial quantities, picture a grocery store for a moment without decimals. How would you describe the price of items less than $1? You could use common fractions (fractions written using the notation we discussed in the previous three chapters).

$\$\frac{4}{10}$ $\$\frac{2}{5}$ $\$\frac{1}{2}$

Most likely, you would write all your fractions with the same denominator to make them easier to compare. Since there are 100 cents in $1, it would make sense to use 100 as the denominator.

$\$\frac{40}{100}$ $\$\frac{40}{100}$ $\$\frac{50}{100}$

Now it is easier to compare the cost of each item. Decimals further simplify expressing these costs by letting the place, or location, of the number show its denominator, saving us from having to write "100" every time.

$ 0.40 $ 0.40 $ 0.50

The word *decimal* comes from the Latin root *decem*, which as an adjective means "based on 10."[68] In our place value system, each place is 10 times the previous one. Our place value system is a decimal system.

---

68. *The American Heritage Dictionary of the English Language*, 1980 New College Edition, s.v. "decimal."

Decimals are an extension of our place value system to include partial quantities. Just as the place of a whole number tells us if it represents ones, tens, or hundreds, the place of a decimal number tells us if it represents a fraction of 10, 100, 1,000, and so forth. Each place to the left in our decimal system is worth 10 times the previous place, and each place to the right is worth $\frac{1}{10}$ of the previous place. We use a dot, or decimal point, to separate the whole numbers from the partial ones.

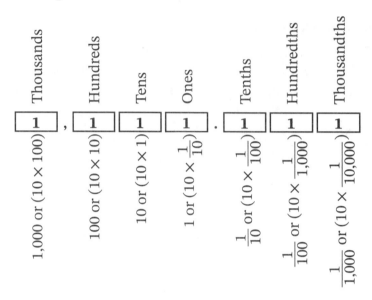

Keeping Perspective

Decimals, like fractions and whole numbers, describe quantities. We are only able to describe quantities using decimals because God gave us this ability. Mice do not do math, but man can because God created man in His image. Isn't it wonderful to think God designed us differently than animals and gave us the ability to fellowship with Him?

The account of King Nebuchadnezzar (Daniel 4) really brings this point home. King Nebuchadnezzar boasted to himself about Babylon, taking credit for building such a great empire. God humbled him and made him like the beast of the earth. God was showing him *he could not even think or function apart from God's enabling*. We are utterly and completely dependent on God for *everything*, including the ability to think and name quantities.

A Look at History

Decimals, or decimal notation, were gradually developed and adopted to help simplify various aspects of math. Adopting this new notation took time.

When decimal notation was first gaining popularity, mathematicians used many different ways to distinguish between whole and fractional amounts. It was common for each mathematician to adopt his own symbols. Below are just a few ways people have used to symbolize $2\frac{16}{100}$, or 2.16. Some of them are quite interesting — notice how one mathematician even used our current equal sign as a way of separating whole numbers from fractional amounts.[69]

2⓪1①6②	216②	2.1̇.6̈.	2,1̸6̸⁄⁄	2:16
2⌊16	2=16	2▲16	2¹⁶	2'16

Over the years, mathematicians have tried to standardize notation. After all, it is a lot easier not to have to learn a whole new set of symbols for every math book you read. In America, we use a dot to the left of the fractional amount to let us know it is written using decimal notation. In other countries, however, a comma or other variations are used instead of the dot.

History reminds us our current notation is only one way to describe quantities on paper! Far from being some sort of independent truth, decimals aid us in exploring God's creation and serving Him.

Working with Decimals

All the rules about adding, subtracting, multiplying, and dividing decimals help us describe on paper the consistency God holds in place. Let us quickly illustrate this point with each of the different operations.

Adding and Subtracting

One of the biggest advantages to decimals is that, since they are based on 10 just as the rest of our place value system, we can use the same basic methods to work with them as we do whole numbers!

4.12 + 2.6	5.98 − 1.5
4.12	5.98
+ 2.60	− 1.50
6.75	4.48

To add and subtract, then, we can follow the same basic method as whole numbers, only to help keep the digits in the correct place, it is often helpful to add zeros to the right of the number, as we did with 2.6 and 1.5 in the examples. Adding (or removing) a zero to the right of the last number does not change the value of the fraction; it actually multiplies (or divides) the fraction by $\frac{10}{10}$, which is worth 1.

---

69. Symbols based on those presented in Cajori, *History of Mathematical Notations*, 1:314–335.

Top row, left to right: Belgian Simon Stevin, 1585; Wilhelm von Kalcheim, 1629; J.H. Beyer, 1603; John Napier, 1617; Richard Balam, 1653.

Second row, left to right: William Oughtred, 1631; Johann Caramuel, 1670; "other writers," such as A.F. Vallin, 1889; one of many methods referenced by Samuel Jeake (1696) as being in use; another one referenced by Samuel Jeake.

$$0.6 = \frac{6}{10}$$

$$\frac{6}{10} \times \frac{10}{10} = \frac{60}{100} = 0.60$$

$$0.50 = \frac{50}{100}$$

$$\frac{50}{100} \div \frac{10}{10} = \frac{5}{10} = 0.5$$

Figure 9: Adding or Subtracting a Zero to the Right of a Decimal Does Not Change the Value

Multiplication

To multiply decimals, we ignore the decimal point and follow the same multiplication process used with whole numbers. After multiplying, we count the number of digits to the right of the decimal points in the numbers being multiplied, and add a decimal point in the answer so as to keep the same number of total digits to the right of the decimal point. While this process sounds a bit arbitrary, in reality it reduces a process of multiplying and dividing by 10 to a simple "rule" or method. To understand this, let us think through how to multiply 7.5×5 without following this "rule."

We need to first find a way to remove the decimal point. Since our place value system is based on 10, multiplying by 10 or a power of 10 is just a matter of moving the decimal point to the right, which increases the value of each digit. If we multiply 7.5 by 10, this would remove the decimal point, leaving us 75.

$7.5 \cdot 10 = 7\,5.$ → Decimal point moved to the right.

(Ones, Tenths / Tens, Ones)

We can now find the answer to 75×5 using the rules we have already learned. After we have finished, we can divide by 10 again to put the decimal point back in the correct place.

	Multiply to remove the decimal.	Solve.	Divide to add decimal back.
7.5 × 5	7.5 × 5 ($7.5 \times 10 = 75$)	2 75 × 5 375	2 75 × 5 37.5 ($375 \div 10 = 37.5$)

The rule about counting the digits reduces multiplying and dividing by 10 or a power of 10 to a mechanical process we do not even have to think about. But it is really a way of describing multiplying by 10 as many times as necessary to remove the decimal point, and dividing by 10 the same number of times. Since we are both multiplying and dividing by the same number, the multiplication and division will cancel each other out and not affect the final result.

Division

The rule for dividing decimals is similar to the rule for multiplying them. To divide decimals, we first remove the decimal point from the number we are dividing by (the divisor), moving the decimal point in the number we are dividing (the dividend) the same amount. Then we divide as normal without having to worry about the decimals. Note that in moving the decimal point, we may have to add 0s; here, we added a 0 to 50 because we moved the decimal point over to the right.[70]

$$2.5\overline{)50} \qquad\qquad 25.\overline{)500.} \begin{array}{r} 20. \\ -50 \\ \hline 00 \\ -0 \\ \hline 0 \end{array}$$

When we move the decimal points in both numbers, we are really multiplying both numbers by 10 or a power of 10. Multiplying by 10 (or a power of 10) does not affect the answer because we are multiplying both numbers by the same amount. This is easiest to see using fractions. Remember, a fraction represents division. When we multiply both the numerator and the denominator by 10, it is the same thing as multiplying the equation by 1, since $10 \div 10 = 1$.

$$\frac{2.5}{50} \times \frac{10}{10} = \frac{25}{500}$$

Converting Numbers from Common Fractions to Decimals

Common fractions can easily be converted to decimals by dividing the numerator by the denominator. Why? If you will recall from our exploration of fractions, fractions represent division. The fraction "$\frac{4}{25}$" is another way of writing $4 \div 25$. So to convert common fractions to decimals, it would make sense that all we would need to do is complete the division.

$$25\overline{)4.0} \begin{array}{r} 0.16 \\ -2\,5 \\ \hline 1\,50 \\ -1\,50 \\ \hline 0 \end{array}$$

Conclusion

Like common fractions, decimals are a useful way of recording and working with partial quantities. The "rules" used with decimals work because they describe consistencies God created and sustains! The rules are reliable because God is reliable and never changing. Their reliability reveals God's never-changing character. Decimals, and the rules for

---

70. We can think of 50 as 50.0 Moving the decimal to the right would then require adding a 0, giving us 500.

DECIMALS 129

working with decimals, serve as a useful tool we can use while depending on God and joyfully doing the work He has given us.

TEACHING SUGGESTIONS AND IDEAS

Objective: *To help your child view decimals as a useful way of recording quantities; to teach him to easily recognize and record decimal quantities.*

Specific Points to Communicate:

- *Decimals are another useful way to express partial quantities.*
- *Rules work because they describe what happens in real life.*
- *We can worship and praise God while using decimals in a variety of tasks.*

Because our money system uses decimals ($1.34, $20.45, etc.), it provides a good basis for helping children grasp the concept. Consider pulling out some change and dollar bills from your purse and explaining how our money system works (100 pennies equal $1, etc.). Then show your child how we use decimals to represent anything less than $1. Things less than $1 are partial quantities (parts of a dollar).

At some point after the child is comfortable working with decimals and money, you can work on understanding how fractions can convert to decimals and teach more of the theory behind decimals. Please see the example section for ideas. Because decimals are extensions of our place value system, it is a good idea to review the concept of place value (see **PLACE VALUE**) when first presenting the theory behind decimals.

Once your child knows how to write decimals, he will also explore how to add, subtract, multiply, and divide using decimals. Please see the addition, subtraction, multiplication, and division chapters for ideas on presenting these operations — the same principles apply to decimal numbers as to whole numbers. When presenting rules on where to put the decimal point when multiplying or dividing decimals, help the child understand *why* we put the decimal where we do. It is not an arbitrary rule!

When you teach rules, try to have your child come to the rule himself. Instead of having him memorize the rule as a mysterious fact, you want him to learn to think through problems and use mathematics as a useful tool — and to see each rule as a way of describing how God causes quantities to operate. See the explorations of decimal rules earlier in the chapter for ideas of how you could do this.

As your child begins working with decimals, you may want to add more practical examples to your textbook. It is easy to teach mechanics and not really let your child use what he is learning. So as you teach decimals, look for ways to let your child apply this tool! Since we use decimals frequently, you should not have to look very hard. The "Ideas" section lists a few applications to get you started. Many of these applications provide a wonderful opportunity to examine biblical principles on both work and money. Taking advantage of this discipleship opportunity can help show your child that biblical principles and motivations should govern how we apply math.

Example

Your textbook will most likely try to explain to some degree how the decimal system records partial quantities. Some, such as the one below, do a better job at this than others. The one below starts by explaining how 3 dimes could be written as $3\left(\frac{1}{10}\right)$ of a dollar ($3\left(\frac{1}{10}\right)$ is another way of writing $3 \times \frac{1}{10}$; each dime is $\frac{1}{10}$ of a dollar, so 3 dimes would be $3 \times \frac{1}{10}$), then introduces decimals and has the child write various quantities using different notations.

> *Another way to write $3(\frac{1}{10})$ is .3 and is read 3 tenths.*
>
> *This dot in front of the three tells you that it is a fractional part of the whole. In this case, three tenths of a dollar. The dot is called a DECIMAL POINT. It is sometimes read* point three. *The number to the right of the decimal point is a fraction part. Writing a number in this way is called DECIMAL NOTATION.*[71]

Notice how the presentation above clearly introduces decimals as another way of describing quantities and gives the child opportunities to describe the same quantity different ways. Not all presentations do such a good job, and not all children will grasp these points instantly.

One way you could help present decimals would be to put prices on items around the house using common fractions ($\frac{4}{10}$, $\frac{2}{5}$, etc.). Then guide your child into seeing how writing partial quantities with the same denominator would make it easier to compare the prices, and writing them using place value would be easier still! You could also show your child some of the other symbols besides the decimal point used to separate whole numbers from partial numbers (see the chart earlier in the chapter) and go over the idea of place value and how decimals extend the system for partial quantities.

Once the initial concept of decimals is mastered, students learn a lot of rules on working with decimals. It is all too easy to spoon feed these rules and expect your child to memorize them without really demonstrating how these rules describe real-life principles God holds in place. Try using the explanations in the beginning of the chapter to help your child understand the "rules."

Ideas

- **Form decimal numbers on an abacus, and have your child read and write them.** See Appendix C for instructions on making an abacus. You can easily make your abacus represent decimals by viewing the last two or three rows of beads as decimal places (you might string those rows with different colored beads, or draw a decimal point above them). Have fun forming numbers and having your

---

71. Quine, *Making Math Meaningful Level 5*, 239.

child identify them — or let him form the numbers and see if you can read them correctly!

- **Review decimals using money.** Pull out some bills and coins from your purse and have your child practice recording quantities. When you go to the store, use sale signs and price tags to reinforce decimals. Ask your child to explain what the numbers represent. (Example: $1.99 represents 1 dollar and 99 cents. We know the 99 means cents instead of dollars because of its location, or place, after the decimal.) As your child learns how to add, subtract, multiply, and divide decimal numbers, let him pay bills, balance a checkbook, and compare prices.

  While teaching your child how to work with money, take advantage of the opportunity to explore what the Bible says about money. You might start by examining the love of money together. The Bible has a lot to say about the love of money, which can take on so many different forms. There is a snare to spend it too freely, and another one to hoard it. Here are some scriptures you might want to look up with your child: Ecclesiastes 5:10, Matthew 6:24, Luke 16:14, 1 Timothy 6:10, 2 Timothy 3:2, and Hebrews 13:5.

- **Have your child practice filling in pretend or real checks.** Writing a check involves writing decimal quantities using symbols ($15.75) and words (fifteen dollars and 75 cents). Searching the Internet for "pretend checks" should turn up some templates you can print and use. Alternately, you could use one of your own checks, voiding it afterwards.

- **Review previous ideas substituting numbers with decimals.** Since decimals help us add, subtract, multiply, and divide partial quantities easily, reworking some of the ideas listed under those concepts can help show the usefulness of decimals.

- **Look at different symbols used to express decimals.** Show your child the chart of other symbols given earlier in this chapter. You could even have him write numbers using the various notations. Seeing other ways to represent decimals helps us view our current notation as one useful way to express a quantity.

- **Have your child plan a trip.** If your family is going on a trip, let your child estimate all the costs involved using the trip planning worksheets on pages 181 and 182. If you do not have any trips planned, invent a pretend one. You could have your child pull out a map (or use an Internet service) to figure out how far you would have to drive to a particular destination. Have him compare routes, then add up various costs (hotels, etc.). He can use the worksheets as a guide.

- **Have your child plan a party or event.** Multiplication, division, and decimals come in handy when planning a party or event. For example, you might need to calculate the cost of different supplies as well as the number of supplies (packages of plates, pizza, cookies, etc.) to purchase.

- **Have your child use decimals to explore God's creation.** You can use basically any aspect of science to give your child an opportunity to use decimals to explore

God's creation. The "Decimals — Math and Our Bodies" worksheet on page 183 will get you started.

◆ **Set up a "store" with your child.** If you want to have some fun and teach your child to use decimals as a useful tool at the same time, try opening up a "store." Simply clear off a table and pull out a handful of objects your child can pretend to sell. Have your child price each object using numbers with decimals such as $1.99, $2.25, etc. (Suggestion: Write the price of each object on a piece of masking tape so you can pull the "price tag" off when you have finished pretending.) Make play money or use real money that will be returned after the exercise to buy items at your child's "store."

Encourage your child to use what he has learned about decimals to total the objects and give the correct amount of change. You can have him do this with or without a calculator.

◆ **Take a look at how various occupations use the decimal notation.** Looking at how occupations — particularly those in which your child has an interest — require math can be very helpful in demonstrating math's usefulness. No matter what occupation your child pursues in life, he will need to know how to work with decimals, even if just to handle the money side of the business.

» **Music Teachers** — As a piano teacher, I used decimals when calculating the price of music purchased for students.

» **Printers and Graphic Artists** — Every time I prepare a product to take to the printer, I am reminded that even non-math-oriented occupations use math. Printers often deal with converting fractions to decimals when they format projects. Suppose you were formatting a document on the computer and wanted an $\frac{1}{8}$ inch extra margin on the right side of your paper to leave room for comb binding. Since the computer does not recognize fractions, you would need to rewrite $\frac{1}{8}$ as a decimal (0.125) in order to tell the program to leave the appropriate space.

In a similar manner, graphic artists often use decimals to specify the exact dimensions or location of a picture or project.

» **Farmers** — Farmers use decimals to calculate their profits and costs (400 bales of hay at $1.99 a bale).

» **Managers** — Managers use decimals to track their profits and sales.

» **Clerks** — Clerks use decimals to give customers back the correct change.

» **Sports** — Sport commentators, coaches, and players (and fans — but being a fan is not technically an occupation ☺) use decimals to keep track of statistics like batting averages.

» **Scientists** — Scientists use decimals in nearly every aspect of their occupation. Decimals aid in measuring quantities for experiments, recording precise measurements and results, and solving a myriad of math problems which arise when exploring creation.

Make a game out of finding different ways various professions use decimals. As you notice how decimals apply in various professions, take a look at what the Bible has to say about work itself and point out that math is useful in the work God's given us to do.

From a biblical perspective, work is a privilege, not something to be dreaded. You could discuss how our kingdom and treasure is not of this world (John 18:36; Matthew 6:18–19; Philippians 3:19–21; and 2 Timothy 2:1–7), yet at the same time God calls us to live here and serve Him wherever He has called us (Ecclesiastes 2:24; Acts 18:3; and Colossians 3:17).

PARTING NOTE

Decimals are literally used all the time and in every occupation. What a blessing it is that in each aspect of our lives, wherever we are and whatever we are doing, we have the privilege of seeking God, surrendering to Him, and resting in His love.

Seek the LORD, and his strength: seek his face evermore.

PSALM 105:4 (KJV)

Percents

25%   87%   12%

Percents appear neutral. But what do they really represent? How are they used? Answering these questions can "reveal" percents, showing us yet another useful tool for expressing quantities.

What Are Percents and What Do They Represent?

Percents, like fractions and decimals, are another way to refer to quantities as parts or portions of other quantities. The word *percent* is "short for Latin *per centum*, by the hundred." *Per* actually means "by" and *centum* means "hundred."[72] Thus "25 percent" means "25 per 100," or $\frac{25}{100}$.

To save ourselves from writing the whole word "percent" each time, we typically abbreviate it in some way. *Percent,* or *per cent,* has been expressed *p cento, per c̊, per °̸*, and other various forms.[73] Today, these are typically shortened to just %. This symbol is a shorthand way to let us know a number is representing a portion of a hundred. Notice how there are two 0s in the % sign, just as there are two 0s in 100.

$$25\% = 25 \text{ per } 100 \text{ or } \frac{25}{100} \text{ or } 0.25$$

Since a percent is a way of describing a portion of 100, all we have to do to convert decimals to percents is to multiply the decimal by 100 and add the percent sign. Percent signs, like fraction signs, represent division. They mean "per 100," or divided by 100. So to convert back from a percent to a decimal, we simply divide by 100.

Do you see how understanding what the word *percent* means helps us understand *how* percents record quantities, equipping us to use them more effectively?

0.40	40%	$\frac{40}{100}$
0.25	25%	$\frac{25}{100}$
0.30	30%	$\frac{30}{100}$
0.23	23%	$\frac{23}{100}$
0.125	12.5%	$\frac{12.5}{100}$

How Are Percents Used?

In his book *Rudiments of Arithmetic*, James D. Nickel offered this insight into the purpose of percents, "The idea of percent was invented as a means to write all fractions with a single common denominator; i.e., 100. 'One hundred' serves as a standard of comparison. The value of this standard is that it enables people to convert any fraction or decimal fraction into a form in which it can easily be compared with other fractional quantities."[74]

---

72. *The American Heritage Dictionary of the English Language,* 1980 New College Edition, s.v. "per cent."
73. Smith, *History of Mathematics,* 2:247–250.
74. Nickel, *Rudiments of Arithmetic,* 516.

In other words, percents give us an easy way to compare quantities. Because of this, they prove useful in lots of situations.

For centuries, percents have been used when working with interest rates and money. They provide an easy way to represent how a quantity will increase (it will grow by 2%, etc.) or what portion of a quantity is needed (a 5% sales tax or late fee).

Percents are also used to represent how a certain opinion or quantity relates to the whole — what percent of people plan to vote for which candidate, approve of the current President, etc. Saying 60% of the people must agree in order to pass a law is the same thing as saying 60 out of every 100 people must agree. Notice this does not tell us how many people are voting. It only tells us what portion of those people have to vote to get the law passed.

Of course, all these numbers could also be represented by fractions or decimals. Percents just provide a different notation that makes it easier to compare because, rather than looking at the exact quantities involved, we are able to look at everything as a portion of the same number (100).

Percents, like decimals and any other math concept, can be used for good and for evil. Unfortunately, percents are frequently presented in a misleading way. One international politics book I read used percents and a graph of percentages to represent who voted for which candidate in an election. Unfortunately, the percentages did not add up to 100%, nor did they support the claim made. Percents were being used deceptively.

For another example, suppose a business claims their new vitamin solution is 99% effective at keeping people healthy. If the business only tested their solution on people who do not normally catch colds, then saying the solution is 99% effective is deceptive.

As we see percents used in various ways, we need to remember we live in a fallen world. The data behind a percent does not always support the claim, so beware. The Bible teaches us God is the only one we can trust completely.

> ...let God be true, but every man a liar.
>
> ROMANS 3:4B (KJV)

Conclusion

Fractions, decimals, percents — they are all useful ways of recording partial quantities. Just as different names are used to refer to animals [a bird can be referred to as an oiseau (French), pájaro (Spanish), vogel (German), птица (Russian), etc.],[75] quantities can be expressed different ways.

The Bible gives us a foundation for understanding why we can use percents and guides us as we use them. We can use percents because of the ability God gave us. As we use percents, we should seek to use them honestly, at the same time not believing every claim we read, knowing we live in a fallen world where dishonesty and mistakes abound.

---

75. Names based on those given at www.wordreference.com; picture of Russian name comes from the same site.

TEACHING SUGGESTIONS AND IDEAS

Objective: *To help your child view percents as a useful way to record quantities and learn how to use percents in his own life.*

Specific Points to Communicate:

- *Percents are another useful way to represent quantities.*
- *Percents can be used in a misleading way; we need to learn to test what we read, as men are fallen.*

Math is a cumulative subject. As you present percents, you will be building on the foundations you have already laid when teaching fractions and decimals, introducing your child to yet another option for viewing and recording partial quantities.

Example

Below is how one textbook presents percents.

> *We use fractions and percents to describe parts of a whole. We often use the symbol % to stand for* percent. *If your score on a test is 100%, that means you have every answer right since 100% means the whole thing...*[76]

There are several good characteristics about the presentation above. For one, it says "we often use the symbol %," rather than making it sound as if we have to use this symbol. For another, it makes it clear that percents, like fractions, describe parts of a whole.

Despite these good attributes, the presentation does not really leave the child with a very clear understanding of what the percent sign is all about. Why does 100% represent the whole thing? Is the percent sign an arbitrary symbol? Why do we use it?

You could "reveal" percents by explaining the meaning of percent (divided by 100), and explaining how using the same denominator (100) helps us compare quantities.

Consider also bringing in some real-life examples. Most textbooks are filled with word problems that, while they can help the child learn to apply and grasp percents, often leave the child wondering, "Who cares?" Take the following word problem for example.

> *If 8 of the 20 students are boys, what percent of the students are boys?*[77]

This word problem does not mean much to a child. Why should he really care what percent of the students are boys? Word problems with more real-life importance can add much more interest.

One great way to really show the usefulness of percents (and how they can be misrepresented!) is to pull out a newspaper and have your child look through it for percents. You could even make a game out of it! Along the way, you could examine some of the different uses and talk about the assumptions behind the conclusions. You have now shown your child how useful percents can be, and taught him a bit of discernment in the process.

---

76. Hake and Saxon, *Math 54: An Incremental Development*, 222.
77. Hake and Saxon, *Math 65: An Incremental Development*, 443.

Ideas

- **Open a newspaper or magazine and see how many uses of percents you can spot.** This is a great way to show your child percents are used all the time!

- **Open a "store" and run various sales.** See instructions in the "Ideas" section of the decimal chapter for opening a store. Have your child run percent-off sales (20% off, 40% off, etc.), figuring out how much to charge for various items you or other family members want to "purchase." For a real challenge, have your child figure out how much he would make on various items if he put them on sale based on pretend "purchase costs." Be sure he includes sales tax.

- **Highlight various ways percents apply in everyday life.**
  - » **Food** — The nutritional information on the back of food boxes is listed as a percentage of the daily recommended intake.
  - » **Shopping** — Percents prove useful at the store; sales tax, as well as many sales, are expressed as percentages.
  - » **Fertilizer** — The numbers on the front of a fertilizer bag (such as 5-10-5-0) list the percent of nitrogen, phosphorus, potassium, and sulfur in the fertilizer.[78]
  - » **Restaurant** — Decimals and percentages both come in handy when leaving a tip.

- **Explore pie graphs with your child.** Pie graphs pictorially compare data by illustrating what percent of the whole each expense, opinion, etc. represents. Searching the Internet for "pie graphs" should turn up lots of examples.

- **Teach your child about interest rates.** Since interest rates are expressed as percents, they provide a wonderful example of how percents serve as a real-life tool. If you need help understanding them, your library should have a practical math book you can use to help you understand how interest rates operate.

PARTING NOTE

Don't you love God's creativity? He did not just give us the ability to think of one way to record partial quantities — he created us capable of using many different valuable methods for recording His creation. Man's creativity is but a shadow of God's!

---

78. Sometimes the fourth number (sulfur) is omitted. Fertilizer information found on the GardenLine website. Sara Williams, "Fertilizer: The Basics," GardenLine website, University of Saskatchewan Extension Division, the Department of Plant Sciences, and the Provincial Government, http://gardenline.usask.ca/misc/fertiliz.html (accessed August 5, 2009).

Ratios and Proportions

Ratio is a word used to describe a specific way of comparing two quantities — if we sold 7 slices of apple pie and 1 slice of rhubarb pie during the morning coffee break, we would say the ratio between what we sold was 7 to 1 or 7:1.

Proportion is a word used to refer to two equal ratios. For example, 3:6 and 4:8 form a proportion — the first numbers (3 and 4) are both half the second numbers (6 and 8).

But what is the point of learning ratios and proportions? Is it just a bunch of empty head knowledge or meaningless bookwork?

Absolutely not! Both ratios and proportions are fancy names for specific ways of comparing quantities. As such, they prove useful in many situations. Let us "reveal" ratios and proportions by taking a closer look at what they are and how they are used. As we do, we will see that, like the rest of math, ratios and proportions assist us with the tasks before us — and reveal more of God's remarkable design.

Understanding Ratios

Suppose a recipe called for 3 cups of flour and 1 cup of sugar.

3 cups flour   1 cup sugar

We could use a ratio to compare these quantities. My dictionary defines a ratio as, "The relative size of two quantities expressed as the quotient of one divided by the other."[79] In other words, *ratio* is a fancy name for using division to compare quantities! So the ratio of flour to sugar would be the quotient (result) of 3 divided by 1, and the ratio of sugar to flour would be the quotient (result) of 1 divided by 3.

Since fractions are a way to represent division, we could use fractions to represent ratios. It is also commonly accepted to use a colon to represent ratios. There is nothing special about a colon necessarily, except that it is a common convention. It is a lot easier to read math books if every one uses the same convention. In the chart you can see three different conventions to represent the ratio, or relationship, between the cups of flour and the cup of sugar.

Ratio = relationship found by dividing one quantity by the other

Ratio of flour to sugar: 3:1 or $\frac{3}{1}$ or $3 \div 1$ | Ratio of sugar to flour: 1:3 or $\frac{1}{3}$ or $1 \div 3$

---

[79]. *The American Heritage Dictionary of the English Language*, 1980 New College Edition, s.v. "ratio."

3:1, $\frac{3}{1}$, 1:3, and $\frac{1}{3}$ all represent division. We could also have written these ratios as 3 ÷ 1 and 1 ÷ 3; however, it is more common to work with ratios written as fractions or with colons.

When a fraction represents a ratio, or relationship between two quantities, we read it differently. We would read these ratios as "3 to 1" and "1 to 3." Again, this is a standard convention to make it clear we are talking about 3 compared to 1 (or 1 compared to 3) rather than about a partial quantity.

In short, ratios are a way of comparing quantities.

Understanding Proportions

Notice how the ratio between the height and the length of these rectangles ($\frac{2}{4}$ and $\frac{4}{8}$) both reduce to the same fraction: $\frac{1}{2}$. Even though these rectangles are different sizes, the ratios between their sides are the same. In both rectangles, one side is $\frac{1}{2}$ the size of the other. We would say these rectangles are proportional, and that their ratios form a proportion.

Proportion is a name used to refer to two equal ratios. Both sets of ratios below form proportions — they represent the same relationship between two quantities.

$$\frac{2}{4} = \frac{4}{8} \qquad \frac{2}{3} = \frac{4}{6}$$

Understanding Means, Extremes, and Cross Multiplication

As you may have noticed, mathematicians like to give names to different parts of equations. Names help us communicate clearly. It is a lot easier to say, "Go to Susan's Grocery Store" than to say, "Go to the place where they sell food that is down the street and to the left." In a similar way, it is a lot easier to use specific words to describe parts of equations. Below are the names typically used to refer to different parts of proportions.

$$\text{Extremes} \underset{3}{\overset{2}{\diagup}} \underset{6}{\overset{4}{\diagdown}} \text{ Means}$$

If we multiply the extremes (2 × 6) and the means (3 × 4), their answers equal! It is this way with every proportion. This is neat, but why does it happen? And why does it matter?

Notice how multiplying the extremes and the means multiplies the same numbers as we would if we multiplied the first ratio by the reciprocal (inversion) of the second ratio. Notice also how in each case, we ended up with equal answers (12), which could be represented as a fraction worth 1 (in this case, $\frac{12}{12}$).

Multiplying the means and the extremes

Extremes $\cancel{2 \times 4}$ Means
$\cancel{3 \times 6}$

$2 \times 6 = 12$
$3 \times 4 = 12$

Multiplying by the reciprocal

$$\frac{6}{4} = \text{the reciprocal of } \frac{4}{6}$$

$\frac{2}{3} \times \frac{6}{4}$

$2 \times 6 = 12$
$3 \times 4 = 12$

To see why we end up with a fraction worth 1, we need to back up for a moment and look at what happens when we multiply a fraction by its reciprocal. By multiplying a fraction by its reciprocal, we are reversing the division; hence, we will always end up with a fraction worth 1.

Note: This is easiest to see with a simple fraction, such as $\frac{1}{3}$. Multiply $\frac{1}{3}$ by 3 (or $\frac{3}{1}$, which means the same thing) and you get $\frac{3}{3}$, or 1.

But wait a minute .... in the original problem worked above ($\frac{2}{3} = \frac{4}{6}$), we did not multiply a fraction by its reciprocal. We multiplied the first ratio by the reciprocal of the second. Why did that give us an answer worth 1? Because if the two ratios are equal, which they are in a proportion, then their reciprocals must also be equal! Thus multiplying the first fraction by the second fraction's reciprocal is the same as multiplying the first fraction by its own reciprocal.

Multiplying the extremes and the means (also called *cross multiplication*) is just a shortcut to multiply the first ratio by the inverse of the second. It is a shortcut for describing a property of fractions — a consistency our God holds in place.

$\frac{2}{3} = \frac{4}{6}$ | $\frac{2}{3} \times \frac{6}{4} = \frac{2 \times 6}{3 \times 4} = \frac{12}{12}$ | $3 \times 4 = 12$, $\cancel{2 \times 4}$ over $\cancel{3 \times 6}$, $2 \times 6 = 12$

Original | Multiply by the Reciprocal | Cross Multiplication

Why does it matter? Well, cross multiplication often helps us solve a real-life problem much easier than we could otherwise. Say you were planning a party. According to the package, you need 2 quarts of lemonade for every 8 people. If you expect 96 people to attend the party, how many quarts of lemonade should you make?

Here you know three parts of a proportion, but need to find the fourth quantity: the number of quarts to make.

$$\frac{\text{\# of quarts}}{\text{\# of people}} \quad \frac{2}{8} = \frac{?}{96}$$

RATIOS AND PROPORTIONS

You can use the knowledge that the products of the means and the extremes in a proportion equal to solve this problem.

The first step is to multiply the means by the extremes (cross multiply).

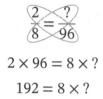

$$2 \times 96 = 8 \times ?$$
$$192 = 8 \times ?$$

The next step is to divide both sides by 8, giving us the number of quarts of lemonade needed: 24. Remember, one way to write division is using a fraction line. Writing the division this way makes it easy to see that the 8s on the right side of the equation cancel out.

$$\frac{192}{8} = \frac{8 \times ?}{8}$$
$$24 = ?$$

We need 24 quarts of lemonade.

In short, cross multiplication is a useful shortcut we can use because of consistencies God holds in place. The way He causes quantities to interact is truly marvelous!

Ratios and Proportions in Action

Ratios and Proportions in Everyday Life

We often find it helpful to compare numbers. Ratios and proportions come in handy while shopping (to compare the cost of different packages), reading maps, making model airplanes, converting measurement units, and more!

Ratios and Proportions in Geometry

Ratios are used extensively in geometry — we can even use ratios to find the height of a tree without leaving the ground!

Notice how we could think of the height of a tree and its shadow as a triangle.

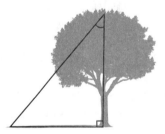

While we cannot typically measure the height of a full-grown tree, we can often measure the length of a tree's shadow. Then, to find the tree's height, all we would need to do is form another smaller triangle proportional to the triangle with the tree.

One way to do this would be to put a stick in the ground and measure its height and the length of its shadow. Because the sun strikes the stick at the same angle it strikes the tree, it forms a proportional triangle. If we measure the stick and its shadow and set up a proportion, we can find the height of the triangle with the tree.

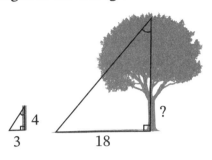

Step 1:
Write the Proportion

$$\frac{4}{3} = \frac{?}{18}$$

Step 2:
Cross Multiply

$$\frac{4}{3} = \frac{?}{18}$$

$4 \times 18 = 3 \times ?$
$72 = 3 \times ?$

Step 3:
Divide Both Sides by 3

$$\frac{72}{3} = \frac{3 \times ?}{3}$$

$24 = ?$
Height of tree = 24

Pretty cool, don't you think? This method for finding the height of a tree could prove valuable if we needed to cut down a tree and want to make sure it will not hit the house when it falls. Not only that, but many believe the Greek mathematician Thales (624–526 B.C.) used this method to find the height of the Great Pyramid.[80]

Ratios and Proportions in Science

Ratios and proportions aid in exploring God's creation, bringing to our attention lots of hidden design and order.

For example, as we compare patterns and aspects of different plants and animals, we find that many different aspects of God's creation express the approximate ratio 1.6180339887. Since we encounter this ratio so much, we have given it a name: the golden ratio. The two spirals in a sunflower always relate mathematically together according to this approximate ratio (one direction containing approximately 1.62 more spirals than the other). The nautilus' sea shell contains a spiral which can be broken down into smaller rectangles that express this same ratio. And the spirals in a pineapple, like those in a sunflower, relate to one another according to this same ratio.

As if that were not interesting enough, many people think rectangles with this ratio, nicknamed "golden rectangles," are visually appealing; hence, you will often find these ratios in art and architecture. The squares in the pictures below are all approximately golden rectangles.

---

80. Thales's story makes an interesting study. If you would like to learn more, searching the Internet for "Thales pyramid" will turn up lots of information.

Isn't it interesting that various and completely unrelated parts of creation express a ratio appealing to the human eye? What a wonderful reminder that the same God created both creation and man! It is He who placed this design everywhere around us and created our minds to appreciate the beauty of His creation.

Not only is the golden ratio visually appealing, but it allows for the most compact spiraling possible, making it the ideal ratio for sunflowers. God sure thought of everything, didn't He? Nothing escapes His notice. Math reveals the detailed care He took with a sunflower's design.

Isn't it comforting to think the same God who created the sunflower, nautilus, and pineapple is watching over every detail of our lives?

Conclusion

Comparing quantities is extremely useful — and ratios and proportions are helpful ways of expressing and referring to number comparisons. We can use these tools in a variety of everyday situations, as well as in more specialized situations (such as to find the height of a tree). Far from being meaningless bookwork, ratios and proportions are tools we can use for God's glory in a variety of settings.

TEACHING SUGGESTIONS AND IDEAS

Objective: *To help your child develop his ability to compare quantities and look at quantities different ways, seeing along the way God's greatness and care.*

Specific Points to Communicate:

- *Ratios and proportions are helpful ways of comparing quantities.*
- *Ratios and proportions aid in seeing the detailed design throughout God's creation.*

By the time your child studies ratios and proportions, he is likely at an age and mathematical maturity to really begin exploring lots of fun math applications. While many ratio presentations and word problems seem dry and meaningless, they certainly do not need to be. Real-life examples and opportunities to apply ratios and proportions abound. It is time to let your child begin using his mathematical toolbox in earnest, applying ratios and proportions to lots of different situations. The following "Example" and "Ideas" sections will get you started.

Example

Notice how the word problem below, while there is nothing necessarily wrong with it, fails to really excite the student about using ratios in a real-life way.

The ratio of humpback whales to killer whales was 2 to 7. If there were 42 killer whales, how many humpbacks were there?[81]

Put yourself in your child's shoes for a moment. If all you day after day you solved problems like this, would you think ratios were very useful, or would you view them as busywork?

On the other hand, what if I were to take you outside and have you find the height of a tree using ratios? Or have you use ratios and proportions to make a model airplane? Or to draw a picture of the Washington Monument? There is nothing like actually using math practically to help students really see it as a useful tool!

There are many, many ways you can have your child apply ratios and proportions — see the following ideas to get started.

Ideas

- **Using your child's interests as a guide, work with him to find examples of real-life ratios.** You could explore the ratio of different objects (picture frames, books, tables, windows), the ratio of ingredients in a recipe, the ratio of wins and losses of a sports team, etc. It might be fun to see how many golden ratios you can find in your own home!

- **Study various applications of ratios within science.** As we look at ratios throughout God's creation, we continually see glimpses of God's thoughtfulness and care. Two worksheets illustrating this are included in the back. The "Ratios & Proportions - The Way Things Grow" worksheet on page 184 uses ratios to examine gracious way God designed our bodies to change over time. The "Ratios & Proportions — Math and Sunflowers" worksheet on page 185 uses ratios to explore the hidden order and design God placed within each sunflower.

  For additional ideas, look for places in your child's science curriculum with comparisons, graphs, or formulas in which ratios have been employed.

- **Explore unit conversion with your child.** Different cultures have different money and measuring systems, and different industries often have specific measuring systems. Within America, we currently use both the metric and the U.S. customary system (also called the English system) in everyday life, as well as a variety of industry-specific systems (in the graphic industry, you find pixels, picas, and points). Converting between different systems requires knowing the ratio, or rate, between them. For example, the ratio between centimeters and inches is $\frac{2.54 \text{ cm}}{1 \text{ in}}$.

---

81. Hake and Saxon, *Math 76*, 249.

Online sites listing how different measurement systems or units compare (quarts to gallons, etc.) abound. You could write this information down as a ratio and have your child practice converting between these different measurement units. You might also want to look up the exchange rate between different monetary units and have your child see how $15 converts to a variety of different currencies — you could even pull out a globe and cover geography at the same time. If you want to really have some fun, have your child pretend to be a world traveler, pick a handful of countries to visit, and make the mental conversions between currencies.

- **Teach your child to use ratios to solve problems.** Ratios can be used in solving everyday problems, whether in the yard or in the kitchen. Assume a bag of mulch says it will cover 2,500 square feet. How many bags do you need to cover 20,000 square feet? Set up a ratio to find out! Or suppose you had to double a batch of cookies. You would want to keep the ratio, or proportion, of the ingredients the same! As you discover everyday applications, share them with your child, giving him an opportunity to apply ratios himself.

- **Have your child make a model airplane/boat/train.** Models are scaled-down versions of objects. The ratios in the model are the same as in the full-sized version. Otherwise, the model would not look like the original.

- **Have your child draw pictures of famous landmarks.** Have him find the dimensions of the landmark in a book or online, then draw the landmark with the same ratios/proportions. For example, you might have him draw a $\frac{1}{200}$th scale version of the Washington Monument. According to the measurements I found, the Washington Monument is 555.427 feet tall, 55.125 feet wide at the base, and 34.458 feet wide at the top.

- **Have your child explore scale drawings.** You can do this many different ways!

  » **Maps** — Maps are basically scale drawings. Having your child find the scale on a map reinforces ratios while teaching important map-reading skills!

  » **Instruction manuals** — Many instruction manuals include a scale drawing or diagram.

  » **Rearranging furniture** — Making a scale drawing of a room and its furniture can help you decide where to place furniture *before* rearranging it. For a fun activity, have your child make a scale drawing of his room and play around on the drawing with different locations for his furniture.

  » **Landscaping** — Making a scale drawing of a garden before planting can help you figure out what to plant where. You might have your child landscape an area of the yard, helping you decide where plants will go.

  » **Construction industry** — Scale drawings are invaluable in the construction industry. Consider taking a trip to the local town hall and asking to see the blueprint on file of your house — it is a scale drawing! Or see if you can set up a visit to a construction site, architecture firm, or civil engineering firm. Let them know you are studying ratios and wondered how they used scale drawings in their industry. You should also be able to find a lot of good material on drafting/house construction online.

PARTING NOTE

As we look at the ratios and proportions all around us, the remarkable design we find continually amazes us. God truly is an amazing architect, even caring for the details of how seeds in a sunflower are arranged. What a reminder to us to live in trust instead of worry!

> *Wherefore, if God so clothe the grass of the field, which to day is, and to morrow is cast into the oven, shall he not much more clothe you, O ye of little faith?*
>
> MATTHEW 6:30 (KJV)

Types of Numbers
(Number Sets)

You have most likely heard terms like *whole numbers*, *real numbers*, *odd numbers*, and *negative numbers*. These number sets are fancy names given to numbers with certain characteristics. Much as Adam used names to describe the animals, we use names to describe and sort the quantities God placed around us.

To better understand the role of number sets, join me in looking at what we mean by organizing numbers into sets and at why we can organize numbers. We will see that number sets, like the other math concepts we have looked at so far, are useful tools we can use because of the way God created and sustains all things.

What Do We Mean by Organizing Numbers into Sets?

In mathematics, a set is "any collection of distinct elements."[82] In a broader sense, a set is "a group of persons or things connected by or collected for their similar appearance, interest, importance, or the like."[83]

Imagine a completely unorganized clothes closet. A closet where, to find a shirt, you would have to sort through a whole stack of different-type clothes.

Organizing avoids this mess. We might organize our closet by color, putting all the blue shirts together and all the red shirts together. Or we might organize it by type, putting all the long sleeve shirts together and all the short sleeve shirts together (or the formal shirts in one spot and the non-formal ones in another). Or we might use a mixture of these systems, or a different system altogether.

Much as we organize shirts by grouping them based on their characteristics, we organize numbers into groups, or sets, based on their characteristics. Some numbers represent whole objects, like one apple or one dollar. Other numbers represent partial objects, like half a cookie. So we might have a set of whole numbers and a set of partial numbers.

Just as a shirt could be categorized different ways (by color, type, etc.), numbers can be (and are!) categorized different ways. The chart shows some commonly used number sets. Notice how some numbers fall in more than one set, and that some sets are subsets of other ones. The different sets categorize numbers based on various characteristics.

Giving names to numbers with certain characteristics helps us easily refer to a specific type number. There is nothing special about the actual names used. Names like *whole numbers*, *counting numbers*, etc. are but descriptive words chosen to categorize the quantities God has placed all around us.

---

82 *The American Heritage Dictionary of the English Language*, 1980 New College Edition, s.v. "set."
83. Ibid.

Note: Many of these sets are not typically introduced until high school; they are included for your reference.

DIFFERENT TYPES OF NUMBERS

COMPLEX NUMBERS		
Includes both real and imaginary numbers.		

REAL NUMBERS		IMAGINARY NUMBERS
Includes both rational and irrational numbers, but not imaginary.[84]		In solving equations, we sometimes come across numbers that cannot be expressed as real numbers, such as the square roots of negative numbers. Since a negative times a negative equals a positive, there is no number times itself that would equal a negative number. So $\sqrt{-1}, \sqrt{-2}, \sqrt{-3}$, etc. cannot really exist. However, we use negative square roots as placeholders to help us solve equations as well as in some branches of mathematics. These numbers are referred to as imaginary numbers.

Rational Numbers — columns: Rational Numbers / Irrational Numbers

Rational Numbers	Irrational Numbers
"Rational numbers" is a fancy name for all numbers except irrational and non-real numbers, including fractions and decimals. They can be expressed as the ratio (i.e., division) of one integer to another.	Mathematicians refer to numbers that 1) never repeat and 2) go on and on for infinity as irrational. Example: 3.14159265... They cannot be expressed as a ratio (i.e., division) of one integer to another.

INTEGERS	NON-INTEGERS
A name for non-fractional numbers. $\{..., -2, -1, 0, 1, 2, ...\}$	

| *Whole/Natural/Counting Numbers* $\{1, 2, 3, ...\}$ *(Some definitions include 0.)*
 Prime Numbers A name for numbers that cannot be divided evenly by any whole number other than themselves and 1.
 Negative Integers $\{..., -3, -2, -1\}$
 Positive Integers $\{1, 2, 3, ...\}$
 Even Integers (Even Numbers) A name for integers that can be evenly divided by 2. $\{..., -4, -2, 2, 4, 6, ...\}$
 Odd Integers (Odd Numbers) A name for integers that cannot be evenly divided by 2. $\{..., -3, -1, 1, 3, ...\}$ | A name for all rational numbers that are not integers. $\{..., \frac{1}{2}, -.3, \frac{1}{2}, .5, ...\}$ |

Likewise, there is nothing special about the notation used to express sets. While sets are typically expressed in brackets {}, we could use a different notation. But a standardized notation helps us communicate ideas and understand what others have written.

Why Can We Organize Numbers?

Biblical principles give us a framework for understanding why we can organize numbers into sets. For one, God's consistency in holding all things together enables us to have confidence certain quantities will consistently have certain properties (for example, even quantities will consistently be divisible by two). If quantities did not operate consistently, grouping numbers (which represent quantities) by their properties would be pretty useless outside a textbook.

Also, if our minds were generated by chance, we would have no reason to suppose we could make logical sets. But the Bible makes it clear that we are not here by chance but

---

[84] Weisstein, Eric W. "Real Number." From *MathWorld* — A Wolfram Web Resource. https://mathworld.wolfram.com/RealNumber.html

were carefully fashioned by a loving Creator. We are able to observe the characteristics of numbers because the God who created all things also created us capable of observing and naming His creation.

Mind-boggling Numbers

As we explore numbers and learn more about their properties, we find out mind-boggling attributes beyond human comprehension. For instance, try as we might, we cannot find a number that does not have a higher number.

The infinite nature of numbers reminds us of our limited knowledge. As James D. Nickel points out, "The infinite nature of the natural numbers has a way of telling man's reason, 'Under certain conditions, you can never know everything there is to know about me.'"[85] Although our understanding is finite, God's understanding is infinite. Psalm 147:5 tells us, "Great is our Lord, and of great power: his understanding is infinite." How foolish it would be not to trust Him!

Patterns in Numbers

Some sets, like odd numbers and even numbers, follow a pattern, or sequence (each number is two greater than the number before it). Looking for sequences (patterns) in both number sets and elsewhere can prove quite helpful. For instance, a pianist might unconsciously memorize the pattern of black keys on the piano (the pattern is 2, 3, 2, 3, etc.) to help him find his way around the keyboard. We memorize the sequence 2, 4, 6, 8, 10, etc. to help us count by 2s. Counting by sequences is very useful when counting money or stacks of objects. Number sequences are also used in other fields, such as cryptology. Some secure networks use randomly generated sequences to verify a person's identity before letting him log in.

Observing number patterns also reveals fascinating order throughout God's creation. A whole journal (*The Fibonacci Quarterly*) has actually been devoted to recording instances where numbers from a sequence known as the Fibonacci sequence are discovered in creation. Adjacent numbers in this sequence form golden ratios (see page 143), which turn out to be both very practical and visually appealing. The Fibonacci sequence helps us notice fascinating order God put throughout His creation.

Why All These Different Types of Numbers?

We have so many different types of numbers because God created a complex universe, and it takes a lot of different types of numbers even to begin to describe it. Complex, irrational, negative numbers — these are all useful ways of describing quantities.

Take negative numbers for example. Negative numbers do not seem to make a lot of sense initially. How can quantities be negative? It is not intuitively obvious how these numbers describe real-life quantities, but they do!

---

85. Nickel, *Rudiments of Arithmetic*, 294.

We need a way to record quantities we owe people. We need a way to record temperatures below zero. Putting a negative sign in front of a number is one way we represent these quantities.

Physicists need a way to record motion in the opposite direction — and guess what they use? Negative numbers! They might label distances north with positive numbers and distances south with negative numbers. They also label electricity flowing one direction with positive numbers and use negative numbers to represent flow in the opposite direction.

Negative numbers, along with other number sets, are a useful system for recording real-life quantities.

Conclusion

We use number sets to refer to numbers with different characteristics. We are able to do this because God created us with the ability of observing and naming His creation. As we explore sets, we see a glimpse of God's greatness. He created such a complex universe — a universe it takes many different types of numbers to describe.

TEACHING SUGGESTIONS AND IDEAS

Objective: *To help your child develop organizational skills and view number sets as a way of organizing numbers.*

Specific Points to Communicate:

- *Number sets are ways of organizing numbers.*
- *We can organize because God created us with this ability.*
- *Different types of numbers help us describe and appreciate God's complex universe.*

From a very early age, children learn to recognize similarities and divide objects into groups, or sets. You might ask your child to organize his toy cars into different sets. He could arrange them by color, by type, by year, or by any other system that makes sense to him.

At some point, you might pull out a piece of paper and show your child the generally accepted conventions for writing sets. Here are three conventions to keep in mind: 1) an element can only appear one time within a set; 2) sets are represented within braces like this: {cars, trucks, vans, SUVs}; and 3) sets can be subsets of other sets (For example, the set {cars, trucks, vans, SUVs} is a subset of the set {vehicles}).

You could also have your child organize his room, a junk drawer, or even a writing assignment (it helps to organize our thoughts before we put them on paper). The skills of observing and grouping items are quite valuable!

When we organize, we observe similarities between objects or ideas and group like objects or ideas together. Look for ways you can help your child grasp these concepts throughout everyday life.

Watch out for wordings such as, "an even number is." This wording by itself, without the proper background, gives the impression that an even number is some sort of independent, self-existent fact. It would be so much more accurate to say, "We call quantities with these characteristics an even number." Help your child understand these fancy names as ways of categorizing the different types of quantities we find in God's creation.

Example

Most presentations of number sets are very factual.

Whole numbers are the counting numbers and zero.

0, 1, 2, 3, 4, 5....[86]

Put yourself in your child's shoes for a moment. From the above presentation alone, what sort of a view of whole numbers would you get? Would you view them as a self-existent *fact*, or as a *name* we use to describe numbers with specific characteristics? "Whole numbers are..." certainly sounds pretty factual. The student is not given any context in which to understand whole numbers as a name used to organize numbers. It would be easy to leave the lesson viewing whole numbers as an independent fact, subtly buying into an independent view of math.

How could you change presentations like this to help your child see whole numbers as a name rather than an independent truth? There are lots of ways! Even a simple change to the wording, such as, "We call the counting numbers and zero whole numbers," can help. You might also ask your child to think of some different names he might use to describe these type numbers, then present the one we typically use today.

And again, it can be helpful (and fun) to put aside the math book and actually do some real-life organizing to demonstrate the whole organizing process and how different number sets organize numbers. The "Ideas" section below lists quite a few suggestions.

Incorporating real-life organizing presents math as a part of life rather than as a textbook exercise. The skill of recognizing similarities, which organizing teaches, also applies to other areas of math (and life).

Ideas

◆ **Have your child organize paperwork, closets, drawers, toys, ideas for a report, animals, plants, or anything else of interest.** I never quite understood the purpose of number sets until I understood these basic principles of organization: 1) like objects go together and 2) objects can be organized based on different characteristics depending on how we plan to reference them. In organizing numbers, then, we put "like numbers" (numbers with similar characteristics) together. And we organize numbers based on different characteristics depending on how we plan to reference them — whole numbers, positive integers, and even numbers all organize numbers based on different characteristics.

---

[86]. Hake and Saxon, *Math 54*, 21.

◆ **Check out the way a library organizes resources.** A "field trip" to your local library can be a great way to show your child the use of grouping into sets.

While at the library, give your child a specific book to find, such as a biography on George Washington. Ask him how he would go about finding that book. He will most likely suggest going over to the computer and finding the call number. Or he might suggest looking in the biography section.

Talk for a few minutes about how libraries organize books. All the biography books are in one section, the reference books in another, and so forth. Ask your child to think of a different way to organize books. The library could have organized books by their type (hardback verses paperback) or by their paper (shiny verses non-shiny). Ask your child to think about why the library might have chosen the system they did (people typically look for a book based on its content, not its type or paper).

Discuss how hard it would be to find a book if the books in the library were not organized. Even a call number would not help if all the books were thrown haphazardly on the shelves!

We can learn two important principles from the organization of libraries:

» Grouping helps us organize information. Having books organized by type helps us easily find them at the library.

» The same item can be organized differently. The books at the library could be organized different ways. Different organizational systems serve different functions.

These basic principles apply to math as well. Grouping numbers into sets helps us organize, and numbers can be grouped into different sets based on different characteristics.

◆ **Research the Fibonacci sequence.** While you are studying numbers is a wonderful time to take a look at the Fibonacci sequence. And there is a lot to look at! Try searching the web for "Fibonacci sequence" and exploring a few of the links. The Book Extras page on ChristianPerspective.net has a link to a blog post I wrote that contains some links on the topic. You may also wish to look at the "Ratios & Proportions — Math and Sunflowers" worksheet on page 185 and the information on the golden ratio on page 143.

◆ **Look for sequences in everyday life.** When you introduce the patterns (sequences) in some number sets (such as odd and even numbers), you could explore other sequences, pointing out how patterns often aid in memorization. For example, memorizing the pattern of whole steps and half steps in a musical scale helps musicians transpose and arrange music. Memorizing the pattern of black keys on a piano helps pianists make their way around the keyboard. Keep your eyes open for patterns around you!

- ◆ **Practice expressing real-life sets using set notation.** When teaching the generally accepted conventions for writing sets, you might have your child organize a drawer, closet, etc., then express the sets he has formed using set notation. You are thus connecting symbols and notation with a way of describing real-life sets.

- ◆ **Study set theory and the applications of sets in probability and statistics.** While this study is beyond the scope of elementary math, it makes a fascinating research project for a high school student reviewing arithmetic. Note: When studying probability, remember to guard against the idea that things happen randomly and "probability is the very guide to life."[87] From the Bible we know God, not chance or probability, is the true Guide to life. He is the One who controls what will and will not happen, not chance. While we can use probability to make logical guesses about the future, we always need to remember God is the One really in charge.

PARTING NOTE

Beauty, power, order, symmetry, infinitude — though these characteristics of mathematics are there for anyone, the Christian sees them in their proper light as reflections of God's attributes. So there is something to be done with content, after all; the teacher periodically identifies these qualities for the students and reminds them that what they are beholding are beams of God's glory.[88]

LARRY ZIMMERMAN

---

87. Joseph Butler. Quoted in Hacking, *Emergence of Probability*, 11.
88. Zimmerman, *Truth & the Transcendent*, 55.

Exponents and Roots

Although for many the terms *exponents* and *roots* bring back scary images of algebra, these concepts are actually nothing to fear. Exponents and roots are simply different notations commonly used to express repeated multiplication and division.

The exponential number 2^8 is a different way of representing 2 multiplied by itself 8 times, or $2 \times 2 \times 2 \times 2 \times 2 \times 2 \times 2 \times 2$.

Because exponents are used extensively in many different areas, it is helpful to have a way to "reverse" them — and this is the purpose of exponential roots. Most of us are familiar mainly with square roots ($\sqrt{4} = 2$) which reverse square numbers. We can form exponential roots, however, as the inverse operation of any exponential number. The examples show a few exponents and roots, their meanings, and their values.

EXPONENTS			ROOTS		
Exponent	*Meaning*	*Value*	*Root*	*Meaning*	*Value*
4^2 "four squared"	4×4	16	$\sqrt{16}$ or $16^{\frac{1}{2}}$ "square root of sixteen"	The number that, multiplied by itself, equals 16.	4
3^3 "three cubed"	$3 \times 3 \times 3$	27	$\sqrt[3]{27}$ or $27^{\frac{1}{3}}$ "cubed root of twenty-seven"	The number that, multiplied by itself 3 times, equal 27.	3

Notice the two different common conventions for expressing roots: the root sign ($\sqrt{}$) and fractional exponents ($\frac{1}{2}$). When the root sign ($\sqrt{}$) is used, a small number is written to the left of it to show the root desired. If there is no number to the left, the $\sqrt{}$ sign is assumed to mean the square root. Another common way to express roots is as fractional exponents ($\frac{1}{2}$). Writing roots as fractional exponents proves extremely helpful in algebra, as it allows us to perform operations on roots without actually computing their values each time.

You may be wondering why we need yet another way to represent multiplication and division. Why do we have to have so many different ways to express the same quantity? Because God created a complex universe, and it takes a lot of different tools to describe it! Our minds cannot easily work with repeated multiplications such as $2 \times 2 \times 2 \times 2 \times 2 \times 2 \times 2 \times 2$, yet we encounter repeated multiplications all the time (such as when looking at the growth of a population or the interest being accrued). Exponents and roots prove an incredibly useful notational shortcut, simplifying many tasks.

To better understand how exponents and roots simplify tasks, let us take a look at some of their uses. First, though, explore with me a snapshot from their history that illustrates the importance of our worldview.

Insight from History

It is common practice to read numbers like "3²" as "3 squared." Why? A look at the Greeks' view of math provides fresh insight on squared numbers.

Greek mathematicians viewed *everything*, including repeated multiplication, in terms of geometry. When the Greeks thought of two of the same number multiplied together, they immediately thought of a square. In a square, the width and height are always the same; thus the area of any square can be found by multiplying the width or height by itself.

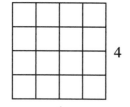

If we were to build a square with sides four units long, the area of our square would end up equaling 4 × 4, or 16. (You can verify this yourself by counting the tiny boxes in the picture.) If we were to build a square with sides 8 units long, the area of the square would be 8 × 8, or 64.

Although we still refer to numbers multiplied by themselves as "square numbers," we do not think about square numbers the same way as the Greeks. Since the Greeks thought about square numbers in terms of geometry, they would never have squared numbers like –4. After all, how can you draw a negative square? That simply would not make sense.

Today, we take square numbers to mean numbers multiplied by themselves, regardless of whether or not those numbers could form a physical square. We square negative numbers and irrational numbers, even though these do not form physical squares.

Why did the Greeks ignore numbers that could not be expressed with geometry, such as $(-4)^2$? It has been suggested that perhaps the Greeks ignored these numbers because they had a clumsy notation system that made working with non-geometry problems difficult. While I am sure this contributed to the situation, I believe there were other reasons too.

Mathematician James Nickel points out in his book *Mathematics: Is God Silent?* that,

> *To the Greeks, Euclid's geometry exemplified the glories of deductive reasoning. They viewed syllogistic thinking as the sole pathway to all truth since this method of thought compels the truthfulness of a conclusion based upon accepted axioms. This overemphasis on deductive geometry impeded the development of mathematics as Richard Courant (1888–1972) and Harold Robbins point out, "For almost two thousand years, the weight of Greek geometrical tradition retarded the inevitable evolution of the number concept and of algebraic manipulation, which later formed the basis of modern science."*[89]

You see, the Greeks did not view math as simply a way of naming or recording God's creation. They viewed math as a creation of man. They placed their faith exclusively in their human reasoning. Since they could not understand negative numbers, they simply ignored them and focused on things they could understand. Their worldview kept them

---

89. Nickel, *Mathematics: Is God Silent?* 54. Internal quote from Richard Courant and Harold Robbins, *What Is Mathematics?* rev. Ian Stewart (London: Oxford University Press, [1941] 1996), xvi.

from exploring areas of math and contributed to their almost exclusive focus on geometry. Our worldview matters!

When we look at math from a biblical worldview, we do not have the same trouble the Greeks had with negative and irrational numbers. The Greeks ran into trouble because they approached math as a man-made structure. Thus things defying human comprehension (like negative and irrational numbers) did not make sense. But when we approach math from a biblical perspective, we recognize the usefulness of negative numbers in describing certain quantities, and we expect to find irrational numbers because God created a complex universe that defies our human understanding. Math makes sense when we realize it rests on God, not human reasoning.

Useful Tools

Exponents and roots simplify and help us solve many, many equations, especially when used hand-in-hand with algebra. Let us explore a few simple examples.

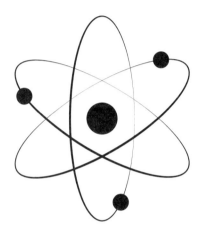

The rest mass of an electron has been calculated at 0.000000000000000000000000000000911 kg — an incredibly small number! Can you imagine trying to write this number often, let alone multiply or divide it with other small numbers? This same quantity can be expressed using exponents as 9.11×10^{-31}, a notation a lot easier to work with!

Exponents give us an easy way to express many growth and decay processes. For example, interest grows in a bank exponentially. To express how much we would have in the bank after 20 years if we started with $500 and had 2% interest compounded annually would be a rather lengthy equation without using exponents.

$$\$500(1 + 0.02)(1 + 0.02)(1 + 0.02)(1 + 0.02)(1 + 0.02)(1 + 0.02)(1 + 0.02)$$
$$(1 + 0.02)(1 + 0.02)(1 + 0.02)(1 + 0.02)(1 + 0.02)(1 + 0.02)(1 + 0.02)(1 + 0.02)$$
$$(1 + 0.02)(1 + 0.02)(1 + 0.02)(1 + 0.02)(1 + 0.02)$$

Using exponents, however, we could rewrite this equation in a more concise form:

$$\$500(1 + 0.02)^{20}$$

Because roots reverse exponents, they prove quite useful when trying to work with more advanced problems. We are able to write general equations to describe the laws by which God has chosen to govern the universe (the law of gravity, acceleration, etc.) using exponents and to rework the equations using roots to find other information we need to know.

Conclusion

Exponents and roots are a different notation for expressing repeated multiplication and division. They prove useful in many different situations, aiding us in both completing tasks and exploring the complexities of God's creation.

TEACHING SUGGESTIONS AND IDEAS

Objective: *To help your child understand how to read and write exponents and roots; to help him apply these tools practically.*

Specific Points to Communicate:
- *Exponents and roots are a different notation for expressing repeated multiplication and division.*
- *Exponents and roots prove useful in a variety of settings.*
- *Worldviews matter.*

Working with exponents and roots can seem like meaningless bookwork, especially in algebra and higher mathematics. The more you help your child see exponents and roots as a useful way of expressing quantities from the beginning, the easier it will be to nurture this perspective in the later grades.

Start by making sure your child understands that exponents and roots are just different ways of writing operations with which he is already familiar (multiplication and division). Then give him opportunities to apply these useful tools. Although many applications require algebra or calculus, the "Ideas" section shares some simpler applications you could explore with your child.

Example

There are many, many different ways to introduce exponents and roots. You can have a lot of fun exploring this new shortcut with your child!

Notice how the following presentation does a good job presenting exponents as a simple way of writing repeated multiplication. It introduces the concept from a real-life situation.

> *In a meeting room, there are 10 rows of 10 chairs. To find the total number of chairs, multiply 10 × 10. Another way to express 10 × 10 is 10^2. In 10^2 the exponent is 2 and the base is 10. The* **exponent** *tells how many times the* **base** *number is used as a factor.*[90]

If your textbook uses a real-life illustration in this way, you could actually draw it and ask your child to picture when he might need to find this information (if putting out chairs for an event). As you teach exponents, clarify that the word *exponent* is but a name and remind your child why we need so many different ways of writing quantities — because of the complexity and grandeur of God's creation!

Another way to present exponents would be to write out a long train of repeated multiplication (such as 2 × 2 × 2 × 2 × 2 × 2 × 2 × 2), asking him for the product. This should give him an appreciation for the simplicity of 2^8!

You could also present exponents through history. Share the history presented earlier in the chapter, reinforcing as you go the importance of a biblical worldview.

Do not forget to have your child apply exponents and roots outside a textbook. You will find several ideas in the next section. Many of the ideas also include suggestions of how you can also show the importance of worldviews at the same time!

Ideas

Note: I have limited the ideas below to those I thought could be more readily understood by students at the upper elementary level. You will encounter many more advanced uses of exponents and roots in upper math courses.

- ◆ **Explore the usefulness of exponents and square roots in measuring and geometry.** Both exponents and square roots come in handy when finding the area of squares and circles (the area of a square is found by squaring its side [i.e., $A = s^2$] and the area of a circle is found by multiplying π by the square of its radius [i.e., $A = \pi r^2$]).

- ◆ **Study interest rates.** Interest rates are great examples of exponents in action! Show your child how to calculate interest, both for bank accounts and car or home loans. You can easily find this information online or in a book at the library.

- ◆ **Explore scientific notation.** Scientific notation uses exponents to express large and small numbers, thereby helping us explore God's creation. Multiplying 0.000000000000000000000000000911 kg (the rest mass of an electron) by

---

90. Calvert School, *Calvert Math, Grade 7*, 6.

0.00000000002 would involve a lot of counting of zeros and could easily lead to mistakes. But multiplying 9.11×10^{-31} by 2×10^{-11} is a rather simple process.

As you teach your child to record tiny numbers (such as the rest mass of an electron) using scientific notation, you can praise God for His greatness and use these glimpses into the intricate design of the universe to build your child's faith. The number 0.000000000000000000000000000911 is incredibly small, isn't it? Yet within even tiny electrons, our God has put remarkable order, variation, and diversity. At one point, men thought they understood the atom; now, we know we have only scratched the surface.

- ◆ **Explore the Pythagorean theorem.** The Pythagorean theorem provides a useful way for finding the missing side of a triangle. It states: "In right-angled triangles the square on the side subtending the right angle [i.e., the hypotenuse] is equal to the squares on the sides containing the right angle."[91] It is typically represented mathematically by $a^2 + b^2 = c^2$. Notice the exponents!

Ready to let your child apply this theorem? Have him solve the "Exponents & Roots — The Pythagorean Theorem at Sea" worksheet on page 186. There are many, many other applications too — the theorem can be used to square corners when laying a foundation and check to see if a bookcase was built squarely!

While studying the Pythagorean theorem, take a quick look at the worldview of its discoverer, Pythagorous. Pythagoras believed everything could be reduced to counting numbers (1, 2, 3, ...). He practically worshiped these numbers as the rulers of the universe.

But Pythagoras's own theorem shows us everything cannot be reduced to counting numbers! To see how, consider a right triangle whose sides are both 1.

» $a^2 + b^2 = c^2$ Pythagorean Theorem

» $1^2 + 1^2 = c^2$ Substituted values.

» $1 + 1 = c^2$ Simplified ($1^2 = 1$).

» $2 = c^2$ Added.

» $\sqrt{2} = c$ Took the square root of both sides.

The hypotenuse (the side across from the right angle) of this triangle is not a counting number, but rather the $\sqrt{2}$, an irrational number beginning 1.414213562... and continuing on and on. Pythagoras's own theorem completely shattered his worldview![92]

---

91. Euclid, *The Thirteen Books of Euclid's Elements*, 349.
92. Nickel, *Mathematics: Is God Silent?* 21-24.

Numbers do not rule the universe. God rules the universe; numbers only help us explore and record the consistent way He rules. The moment we begin enthroning math we, like Pythagoras, are headed for trouble.

- ◆ **Explore simple uses of exponents and roots in science.** As your child begins to learn some basic algebra, he can start exploring some basic uses of exponents and roots within science. We use exponents and roots when working with power, motion (acceleration, etc.), energy, and nearly every other aspect of physics, as well in other fields of science. Open any physics book and you will find lots of equations with exponents and roots. We are constantly seeing the orderly way God holds everything together and using a combination of algebra and exponents and square roots to record that order. As your child learns various laws in science, he will hopefully see how exponents and roots help us express the consistent ways God holds things together. They are truly useful tools.

- ◆ **Explore exponential growth and decay.** Try looking up "exponential growth" online or in an encyclopedia to get ideas. Notice how many different growth processes can be represented using exponents! Using math to observe growth often proves useful. For example, examining population growth can assist a city in planning and making the appropriate number of roads, schools, etc. for its future population.

Exponential growth and decay not only show the usefulness of exponents, but they also provide a great opportunity to talk about the assumptions that go into how we use math. In predicting the population growth of a city, a lot of assumptions could or could not be right. No one knows what will happen in the future — the population could change completely based on a number of unforeseen circumstances. Teach your child to question and really think through conclusions drawn from apparently sound math — assumptions are not always reasonable or accurate.

One good example of this is carbon-14 dating methods, which are based on the exponential decay of carbon. Answers in Genesis has some great resources on their site on the fallacies and false assumptions that go into this method.

PARTING NOTE

As your child progresses to upper-level math concepts, he will study more conventions and symbols like exponents and roots. There are many different symbols and ways of looking at numbers because of the depth and beauty to God's creation. Each new concept you and your child explore — like exponents and roots — helps us describe real-life consistencies God has placed all around us. May our eyes continually turn toward Him in awe and wonder!

Conclusion

Counting, written numbers, comparing and grouping numbers, addition, subtraction, multiplication, division, fractions, decimals, percents, ratios, proportions, number sets, exponents, roots — all the concepts we refer to as "arithmetic" are useful methods to help us count and work with quantities! These methods only work because a faithful, all-powerful God holds all things together predictably.

As you move on from arithmetic to other concepts, I hope you will continue to look at each aspect of math as a method to describe some aspect of God's creation and the consistent way God holds everything in place. More advanced math concepts, while they may seem more complicated at first, are just more powerful, specialized tools for describing the remarkable order God has placed and holds together all around us.

Algebra, geometry, and calculus build upon the principles we have looked at in arithmetic. More advanced concepts work because basic arithmetic works. And basic arithmetic works because God is in charge and faithfully holding all things together! From counting to calculus and beyond, math rests on God.

Not only does math rest on God, but each "tool" in mathematics helps us with the tasks God has given us here on earth and aids us in exploring God's creation, which, though marred by sin, still proclaims God's love and care.

I pray you will continue to have fun "revealing" math with your child, looking beyond the terminology and rules until you see a useful tool we are able to use because of the way God created and sustains this universe.

May God richly bless each one of you!

By His Grace,

Katherine

Math and the Gospel

For the invisible things of him from the creation of the world are clearly seen, being understood by the things that are made, even his eternal power and Godhead; so that they are without excuse:

ROMANS 1:20 (KJV)

Within math, we see a glimpse of God's eternal power and Godhead. His power keeps one plus one consistently equaling two. His infinity makes infinite numbers possible.

Although we might try our hardest, we cannot change math. We can change the symbols or names, but no matter how we refer to or write it, one of something plus one of something else will consistently equal two. Math is not relative. Why? Because God is God and we are not. He, not us, decides how things will be. He set and keeps certain principles in place, and if we want a math that will actually work, we have to conform to those principles.

We tend to change God and His truths to fit our understanding or thinking. We think, "Surely there must be some good in me. God must want me to fix myself up and be a good person. And He would not send me to hell when I tried so hard." But we need to heed what God says in His Word. Just as we cannot change math principles, we cannot change truth. God decides truth, not us.

So what does God say? The Bible teaches that God created a perfect universe, one that was "very good" (Genesis 1:31). There was originally no suffering, sickness, or death. But man rebelled (Genesis 3). This rebellion is sin, and the penalty for sin is death.

The moment the first man, Adam, sinned, death came into the world. Man's relationship to God was no longer perfect. His body began to die. Creation, too, suffered the effects of sin — sickness, suffering, and death had entered the world. One day, Adam would die completely. Unless someone saved him from sin, he would spend eternity separated from his Creator in Hell, a place of perpetual torment.

God, in His loving kindness, already had a plan in place even before He laid the foundation of the world (1 Peter 1:20). Right there in the Garden of Eden, God clothed the first man and woman with garments of skin (Genesis 3:21), foreshadowing the sacrifice of His own Son, Jesus Christ, whom He would send into the world to take the penalty for our sin upon Himself.

We see God's plan of redemption woven throughout the Bible. Over and over again, God showed mankind his need for a savior — and promised to send one. He gave the Law to show mankind how incapable we are of being holy. Just look at the 10 Commandments (Exodus 20) — by their standard, each one of us falls short. As descendents of Adam, we are born in sin, incapable of doing anything good at all. Psalm 51:5 says, "Behold, I was shapen in iniquity; and in sin did my mother conceive me." Jeremiah 13:23 adds, "Can the Ethiopian change his skin, or the leopard his spots? then may ye also do good, that are accustomed to do evil." Even the good things we do are as "filthy rags" in God's sight

(Isaiah 64:6). God wants us to realize how much we need Him to save us. We simply *cannot* save ourselves!

Yet God can save us. The Bible tells us Jesus, God the Son, took on the form of a man (John 1:1-5). Since He was born of a virgin (Matthew 1:23), He was not born with Adam's sinful nature. He was the second perfect man — the second Adam. "For as in Adam all die, even so in Christ shall all be made alive" (1 Corinthians 15:22).

Jesus lived on earth as fully God and fully man, living a perfect life. He then allowed His own creation to beat, mock, and nail him to a tree. Jesus — the very One who upholds all things so consistently that one plus one consistently equals two — humbled himself to die and bear the wrath for all the evil mankind has ever and will ever do (John 19). It is as if He said, "I did that" to every sin ever committed. Can you imagine a king willingly giving his life for one who tried to take his throne? Yet what Jesus did is far greater than even that. The Giver of life and Creator of all died for the very ones who nailed Him to the tree!

Just before He gave up His spirit, Jesus cried out, "It is finished" (John 19:30). He had completed all that was necessary for our salvation. He had paid our penalty.

But the glorious gospel message does not end there. Three days later, Jesus rose again from the dead. In so doing, He conquered death forever (1 Corinthians 15). He now sits at the right hand of the throne of God (Hebrews 12:2). And one day soon, He is coming back again (Revelation 22:20).

God's way of salvation is clear: "Believe on the Lord Jesus Christ, and thou shalt be saved" (Acts 16:31). God offers us Jesus' righteousness and eternal life in exchange for our sin. What an exchange!

If you have not yet handed your life to God and asked Him to credit Jesus' righteousness to you in exchange for your sin, do not delay! Go to Him in prayer right now. What God says, He means and is able to perform, as math's very consistency so clearly reminds us. God's Word tells us both of His wonderful salvation and eternal life for those who trust Him, and of a coming judgment for those who do not.

If you have already trusted in Jesus as your savior, ask yourself, "Am I living in the truth of the gospel?" The Bible tells us we are to live the same way we are saved — admitting our own inability and trusting and clinging to Jesus and His righteousness.

As ye have therefore received Christ Jesus the Lord, so walk ye in him:

COLOSSIANS 2:6 (KJV)

Note: If your heart has been touched by this explanation of the gospel or if you have any questions regarding it, please contact me through the Christian Perspective website. We would be delighted to share more with you on this topic.

Worksheets

Can you imagine reading about a screwdriver for years, but never actually using one? That would be silly, wouldn't it?

It is just as silly for students to read about math concepts for years, but never use them outside a textbook. Math is a tool to record the world around us. Students need to learn how to use it as such!

The worksheets in this section are designed to give students an opportunity to use various concepts practically. I have tried to make the exercises as realistic as possible.

In real life, it often takes more than one tool to get a job done. For example, to make a cake, you need to use several different utensils, such as a beater, bowl, measuring cups, and a cake pan. To build a table, you need to use a variety of tools, such as a saw, hammer, sander, and maybe even a router.

When we use math in real life, we often need to use a variety of different math concepts in order to find the information we need. Walter Sawyer once compared math to a chest of tools.[93] A good workman not only knows how to use each tool, but he also knows when to use each tool. After all, a tool does not do us any good unless we know when to use it (and when we should use a different tool instead). Imagine how frustrating (and embarrassing) it would be to try to use a saw to hammer a nail!

Teach your child when and how to use each concept.

Part of learning a concept correctly is knowing when we should use it, and when we should use another concept instead. In many of the worksheets, students will need to use several math concepts, not just the one mentioned. Please look at the worksheet ahead of time to make sure it is at an appropriate level for your child. Worksheets with a [H] include higher-level skills than those presented in the chapter referencing them.

All the answers are located in the answer key section following the worksheets.

Students can either write their answers on notebook paper, or you can make copies of the worksheets for immediate family members living in the same household.

---

93. Walter W. Sawyer, *Mathematician's Delight* (Harmondsworth Middlesex: Penguin, 1943), 10, quoted in Nickel, *Mathematics: Is God Silent?* 290.

Written Numbers — Numbers Everywhere

Instructions: Listen carefully to this story about a girl named Amy. At the end of the story, there will be a quiz to see if you can remember all the different ways Amy saw written numbers used.

Amy woke up in the morning and looked at her clock. It read 7:00! She knew the 7 stood for seven hours. Seven hours had already passed since the new day had started.

Amy ran downstairs and asked her mom if she could have cereal for breakfast. Her mom said yes. Amy pulled out her favorite box of cereal and the milk carton.

"Do not forget to check the expiration date!" her mom cautioned.

Amy looked. The date on the milk carton was 6/12/2021. Amy knew the 6 stood for the sixth month, the 12 for the twelfth day of the month, and the 2021 for the year. As Amy poured herself a bowl of cereal, she looked at the nutritional information along the side of the box. She was eating 9 grams of sugar.

After breakfast, Amy and her brother and sisters listened while their mother read the Bible to them. Amy noticed little numbers on the page. She knew those numbers marked different verses, making it easier to find specific scriptures.

Later in the day, Amy and her family piled in the car and headed to the grocery store. To get to the grocery store, Amy's mom turned off the highway at Exit 5.

All over the grocery store, Amy noticed little signs with numbers on them. She knew those numbers represented the prices.

After returning from the grocery store, Amy proceeded to the mailbox. As she collected the mail, she noticed numbers on the front of every envelope. She knew those numbers represented places where people lived. They told the postman where to deliver the mail.

"Mom, will you read a book to me?" Amy asked. Amy's mom agreed. They sat down and began reading. But in the middle of the story, Amy's little brother awoke. Her mother needed to stop and take care of him. "It is okay, Amy," she said. "I will finish the story later."

"How will you know where to begin?" Amy asked.

Amy's mom pointed to the little number at the bottom of the page. "I will remember we are on page 8."

While she waited for her mom to return, Amy decided to practice her piano. What was the number 1 doing on top of the first note? Oh, now she remembered. Musicians sometimes use the number 1 to stand for the thumb. If they put a 1 over a note, it means to hit that note with your thumb.

Amy smiled. She had sure seen a lot of written numbers today!

Quiz: Can you remember the different ways Amy saw written numbers used?

Written Numbers — Finding Written Numbers

How many different uses for written numbers can you find in one day? Keep a sharp lookout today for written numbers. Every time you see a different use for written numbers, write it down here. See how many different uses you can find!

Extra Challenge — Get someone else in your family to look for written numbers too. See which of you can find more.

Extra, Extra Challenge — Draw a story or tell your parents a story of how you used numbers today, or come up with a story about how someone might use numbers.

Multiplication: Foundational Concept — Napier's Rods

Make three copies of this page. This will give you 9 blank strips and 3 printed strips. You only need 1 printed strip — you can throw the other 2 away.

WORKSHEETS 173

Multiplication: Foundational Concept — Using Napier's Rods

Use your Napier's rods to help you solve these problems.

1. How much would it cost to buy 8 packages of party favors if each package cost $5?

2. Your favorite cereal is on sale for $3 a box. How much would it cost you to buy 8 boxes?

3. You are landscaping a flower bed. You have been told you need to plant a ground cover every 3 feet. How many feet could you cover with 6 ground covers?

4. You are planning your family vacation, but you need to work within a budget. A certain museum you would like to visit costs $9 a person. If there are 7 people in your family, how much would it cost you to visit the museum?

Extra Challenge — Have fun using your Napier's rods to solve more multiplication problems. You could even make up a few problems of your own!

Division: Foundational Concept — Napier's Rods

Assumes the student has already completed the previous Napier's rods worksheet.

In the last Napier's rods worksheet, you used Napier's rods to help you figure out how much it would cost to buy 8 packages of party favors if each package cost $5. Since you needed to find what $5 + $5 + $5 +$5 + $5 + $5 + $5 + $5 equals, you pulled out your "5" strip and looked at the 8th box where you had already recorded this answer.

But suppose you started with a different set of information. Suppose you knew you only had $40 to spend on party favors and you needed to find out how many $5 packages you can afford to buy. If you bought one, you would still have $40 - $5, or $35. If you bought two, it would cost an additional $5, meaning you would have $35 – $5, or $30 left. You could keep subtracting out $5 until you got down to zero to find out how many packages you could buy, but this process would take a long time!

Instead, you could look at this problem a different way. You want to find out how many times $5 can fit in $40, or what number times 5 would equal 40. This would tell you how many times you could spend $5 ($5 + $5 + $5, etc.) before you reached your $40 limit. If you pull out your "5" strip where you already have recorded what different sets of 5 equal, you can quickly see that 8 sets of 5 equals 40. You now know that you can buy 8 packages of party favors for $40.

We call what you just did division. Division is a name used to describe the process of calculating how many times a quantity would fit in another quantity, or how many times a quantity could divide into another quantity (you could think of division as a multiplication problem backwards). One way to write what you did in finding the number of sets of 5 that fit in 40 is like this: 40 ÷ 5 = 8, which we would read "forty divided by five equals eight."

Use your rods to help you solve the following problems. Notice how they are the reverse of problems 2–4 on the previous Napier's rods worksheet.

1. Your favorite cereal is on sale for $3 a box. If you have $24 to spend, how many boxes can you buy?

2. You are landscaping a flower bed. You have 18 feet worth of area you need to cover. You have been told you need to plant a ground cover every 3 feet. How many ground covers do you need to purchase?

3. You are planning your family vacation, but you need to work within a budget. If there are 7 people in your family, and you only have $63 to spend, how much could you spend per person?

Multiplication: Multi-Digit Operations — Multiplication in Real Life

1. Your car goes 20 miles on one gallon of gas. You have a 9 gallon tank. How far can you go before you run out of gas? *Note: Knowing the answer to questions like this is very important for trip-planning purposes!*

2. You own a company and are considering giving each of your 9 employees a $50 gift certificate as a bonus. How much will this cost you?

3. You are planning the VBS program for your church this year and are expecting 148 children to attend. You would like each child to paint a T-shirt, but you need to make sure the church can afford this. The T-shirts cost $3 each. How much would it cost to give each child a T-shirt?

4. You set aside $5 a week for a project. How much will you have at the end of 1 year? (There are 52 weeks in a year.)

Multiplication: Multi-Digit Operations — Apartment Rental

Note: This worksheet involves adding and multiplying larger numbers.

Fill in the following boxes at the bottom of the sheet based on the data given.

Furnished Apartment
Apartment Rental (Per Month) $2,000
Furniture Storage (Per Month) $600

Unfurnished Apartment with Rented Furniture
Apartment Rental (Per Month) $800
Furniture (Per Month)
  Bedroom $200
  Living Room $100
  Kitchen $200
One-Time Rental Fee $100
Furniture Storage (Per Month) $600

Unfurnished Apartment with Delivered Furniture
Apartment Rental (Per Month) $800
One-Time Delivery Fee $2,600

	Furnished	Unfurnished with Rented Furniture	Unfurnished with Delivered Furniture
Price for First Month			
Recurring Price After First Month			
Price for 4 Months			
Price for 7 Months			
Price for 12 Months			

Multiplication: Multi-Digit Operations — Apartment Rental (Continued)

Answer the following questions based on the data found on the previous page:

1. Which option would be the most economical if you end up staying 1 month?

2. What is the difference between a furnished apartment and an unfurnished apartment with rented furniture for 1 month?

3. What is the most economical option for 7 months? How much more expensive are the other options?

Division: Multi-Digit Operations — Division in Everyday Life (Version A)

We can use division to help us in all sorts of everyday situations!

1. **At work:**
   a. Companies use division and other math concepts to help them determine how much to charge for their products. If a company spent $200 to print 50 books, how much did each book cost them?
   b. In order to cover the additional costs of advertising and maintaining the company and still make money, the company decides they need to charge 8 times the actual printing costs when they sell the book. How much would they charge for the book?

2. **In school:** If a book is 80 pages long, how many pages should you read each day to finish in 4 days?

3. **In the car:** Once, my parents noticed we were not going very many miles before we needed to fill up with gas. We immediately took the car into the mechanic and discovered our car had a major problem.

   You can use division to help you find the approximate miles you were able to drive on one gallon of gas. If you went 90 miles and used 9 gallons, about how far were you able to go on each gallon?

Division: Multi-Digit Operations — Division in Everyday Life (Version B: Decimal Values)

Note: Answers should be given in decimal form.

We can use division to help us in all sorts of everyday situations!

1. **At work:**
   a. Companies use division and other math concepts to help them determine how much to charge for their products. If a company spent $16,000 to print 20,000 books, how much did each book cost them?
   b. In order to cover the additional costs of advertising and maintaining the company and still make money, the company decides they need to charge 8 times the actual printing costs when they sell the book. How much would they charge for the book?

2. **In school:** If a book is 150 pages long, how many pages should you read each day to finish in 20 days?

3. **In the car:** Once, my parents noticed we were not going very many miles before we needed to fill up with gas. We immediately took the car into the mechanic and discovered our car had a major problem.

   You can use division to help you find the approximate miles you were able to drive on one gallon of gas. If you went 297 miles and used 12 gallons, about how far were you able to go on each gallon?

Decimals — Trip Planning, Part 1

Use math to plan a trip! Using the information given, see if you can fill in the blank fields.

Starting City: Topeka, KS   **Ending City:** Sacramento, CA
Miles of Trip: 1,712

Time of Trip if Average 55 MPH rounded to the nearest hour: _____
Time of Trip if Average 65 MPH rounded to the nearest hour: _____

Gas

Gallons Needed:	86
Price Per Gallon:	$ 3.01

Total Gas: _____

Hotel

Room Rate:	$ 85.00
Taxes:	$ 5.00
Other Fees:	$ 1.50

Total Hotel: _____

Food

Day 1	Breakfast	$ 0.00
	Lunch	$ 10.00
	Dinner	$ 15.00
Day 2	Breakfast	$ 4.00
	Lunch	$ 8.00
	Dinner	$ 10.00

Total Food: _____

Other Expenses

Snacks	$ 4.00
Tolls	$ 8.00

Total Other Expenses: _____

Total Cost of Trip: _____

(Find by adding the total gas, hotel, food, and other expenses.)

Note: Mileage is a rough estimate based on Google™ Maps approximation.

Decimals — Trip Planning, Part 2

Try planning your own trip! Pick two cities and find the miles between the cities (use a map or an Internet map). Research the cost of hotels and food to estimate the trip's cost.

Starting City: **Ending City:**

Miles of Trip:

    Time of Trip if Average 55 MPH rounded to the nearest hour: _____

    Time of Trip if Average 65 MPH rounded to the nearest hour: _____

Gas

    Gallons Needed:[94]   _____

    Price Per Gallon: _____

Total Gas: _____

Hotel

    Room Rate: _____

    Taxes: _____

    Other Fees: _____

Total Hotel: _____

Food

Day 1   Breakfast   _____

        Lunch   _____

        Dinner   _____

Day 2   Breakfast   _____

        Lunch   _____

        Dinner   _____

Total Food: _____

Other Expenses[95]

  _____

  _____

Total Other Expenses: _____

Total Cost of Trip: _____

(Find by adding the total gas, hotel, food, and other expenses.)

---

94. You can estimate how many gallons you need by taking the total number of miles you plan to go divided by the average miles your car goes per gallon (find this information in the owner's manual, online at the car manufacturer's site, or by recording your miles and gallons for a couple of fill-ups).
95. Other expenses might include tickets to an attraction, tolls, snacks, etc.

Decimals — Math and Our Bodies

Math can help us appreciate the incredible way God created our bodies. Although no one knows for sure and the number varies based on body size, scientists estimate the average adult has around 60,000 miles of blood vessels in his body[96] (some believe the average is closer to 100,000 miles of blood vessels). You are going to use math to put this in perspective. For both these exercises, assume the average is 60,000 miles of blood vessels and round your answers to the nearest hundredth.

1. The earth's circumference at the equator is approximately 24,900 miles. If we could stack the blood vessels in an average adult end to end, how many times could they go around the earth?

2. According to Infoplease Distance Calculator, the distance from Boston to San Diego is about 2,583.9 miles. If we were driving that distance at 60 miles an hour, it would take approximately 43 hours, assuming we never stopped and could drive as the crow flies (which we cannot, since the roads do not go as directly as a crow could). Use math to compute how many times if stacked end to end an average adult's blood vessels could be stretched over the distance from Boston to San Diego.

Math helps us see the wonder of the bodies God gave us. The fact that 60,000 miles worth of blood vessels fit inside us and all work together to transfer oxygen and substance to our various body parts defies human comprehension. Truly God's wisdom is beyond our understanding! Math reveals the complexity of our body, pointing us to our awesome Creator.

> *I will praise thee; for I am fearfully and wonderfully made: marvellous are thy works; and that my soul knoweth right well.*
> PSALM 139:14 (KJV)

> *Such knowledge is too wonderful for me; it is high, I cannot attain unto it.*
> PSALM 139:6 (KJV)

---

96. Vogel, Steven. *Vital Circuits*, 15-16; *World Book Encyclopedia*, 2:424. Quoted on www.enotes.com.

Ratios & Proportions — The Way Things Grow

Are you ready to go exploring? Good! Put on your exploring hat, and come discover the remarkable way God designed our bodies to grow.

1. Using the centimeter side of a ruler, measure the approximate length of the man's head and the baby's head.

2. Measure the approximate distance in centimeters from the man's neck to the bottom of his shoe (use the foot that is not stepping forward). Now measure the approximate distance from the baby's neck to the bottom of her foot.

3. Set up a ratio between the man's head and his body and the baby's head and his body. What do you notice?

You should have noticed that the ratio, or relationship, between the baby's head and his body was a lot different than that of the man's head to his body. A baby's head is a lot larger in comparison to the rest of his body than a man's.

When we grow, we do not grow proportionally. Our head does not continue to expand at the same rate as our arms and legs. God, in His infinite wisdom and care, designed us to grow in just the right way. He designed our bodies to change proportions, giving us exactly what we needed for different stages of lives. Babies fall a lot, so as babies, we need extra padding on our heads. But as a we grow and become more mature, we no longer need such a large head. In fact, if our head continued to grow proportionally to the rest of our bodies, we would look rather strange and probably be unable to walk. Instead of a larger head, we need larger feet and muscles to support our growing weight. And that is exactly what God gives us!

By looking at the ratios between a baby and adult, we see that God, in His infinite wisdom and care, designed us to grow in just the right way. As our Creator, He knew our need for different proportions as we grow.

Your Father knows what you need before you ask him.

MATTHEW 6:8B (NIV)

Ratios & Proportions — Math and Sunflowers

Today, we are going to explore sunflowers. Did you know that the seeds in a sunflower are arranged in two spirals?

The data below lists five different hypothetical spiral combinations. Explore the relationship, or proportion, between the number of spirals each direction by dividing the number of sprials in one direction (which we'll call Direction 1) by the number of spirals in the other direction (which we'll call Direction 2). Round your answer to the nearest hundredth and write your answer on the lines. What do you notice?

Sunflower A:	**Sunflower B:**	**Sunflower C:**	**Sunflower D:**	**Sunflower E:**
Direction 1: 13 spirals	*Direction 1:* 34 spirals	*Direction 1:* 55 spirals	*Direction 1:* 89 spirals	*Direction 1:* 144 sprials
Direction 2: 8 spirals	*Direction 2:* 21 spirals	*Direciton 2:* 34 spirals	*Direction 2:* 55 spirals	*Direction 2:* 89 spirals
_____	_____	_____	_____	_____

Each of the sunflowers had a different number of spirals. Yet, when we divided the number of spirals going in one direction by the number going in the other direction, we came up with approximately the same answer every time. Regardless of the *number* of spirals in a sunflower, the relationship, or *proportion,* between the two seed spirals did not significantly change.

Guess what? This relationship or proportion between the spirals allows each sunflower to hold the greatest number of seeds possible![97]

In nature, outside influences distort this pattern somewhat — seeds are not all exactly the same size, and external forces such as pressure against other flower buds during seed development may affect the spirals. Nevertheless, math helps us see a general relationship God put into sunflowers that enables them to reproduce quite efficiently. He cares about even the arrangement of seeds. Remember, the God who watches over such a tiny detail in sunflowers is the same God watching over each detail of *your* life.

> *And why take ye thought for raiment? Consider the lilies of the field, how they grow; they toil not, neither do they spin: And yet I say unto you, That even Solomon in all his glory was not arrayed like one of these. Wherefore, if God so clothe the grass of the field, which to day is, and to morrow is cast into the oven, shall he not much more clothe you, O ye of little faith?*
>
> MATTHEW 6:28–30 (KJV)

---

97. The number of spirals each direction are neighboring numbers in the Fibonacci sequence, and the ratio between them approaches the golden ratio (≈1.62). See Nickel, *Mathematics: Is God Silent?* 241.

Exponents & Roots — The Pythagorean Theorem at Sea

Note: Uses the Pythagorean Theorem.

Suppose a sailor needs to travel to a location 5 miles to the south and 3 miles to the west of his current location. If he takes a diagonal line to his destination, how far will he need to travel? Round your answer to the nearest hundredth.

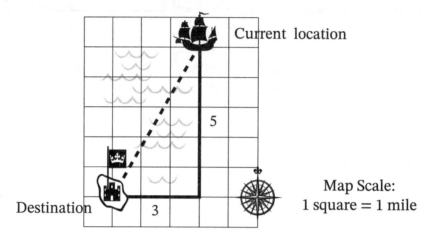

Answer Key

WRITTEN NUMBERS

Numbers Everywhere

Possible Answers: The numbers on the alarm clock, the date on the milk carton, the grams of sugar on the cereal box, the verses in the Bible, the exit number on the highway, the prices at the grocery store, the zip codes and street numbers on the envelopes, the page number in the book, and above music notes.

Finding Written Numbers

Answers will vary. Some possibilities include the ones mentioned in the previous exercise, as well as telephone numbers and the numbers on a phone, apartment numbers, speed-limit signs, scoreboards, numbered questions on a worksheet or test, numerical values on money, rulers, buttons on a copy machine, labels on a map, labels on packages expressing their quantity or weight, radio dials, TV remotes, calculators, computer keyboards, baking temperature gauges, thermometers, timers, and store hours posted in the window.

MULTIPLICATION: FOUNDATIONAL CONCEPT

Using Napier's Rods

Note: You will notice the answers could be obtained from two different strips; the choice of strips does not really matter.

1. $8 \times \$5 = \$40$. It would cost $40.

2. $8 \times \$3 = \$24$. It would cost $24.

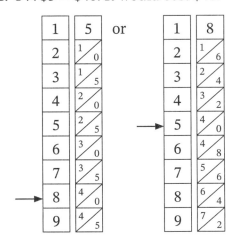
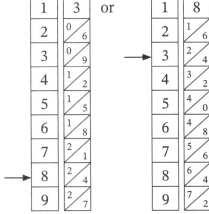

3. 6 × 3 = 18. You could cover 18 feet. 4. 7 × $9 = $63. It would cost $63.

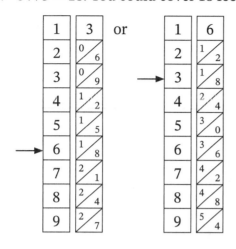
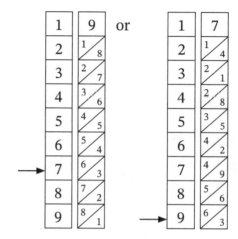

DIVISION: FOUNDATIONAL CONCEPT
Napier's Rods

1. $24 ÷ $3 = 8. You can buy 8 boxes. 2. 18 ÷ 3 = 6. You need 6 ground covers.

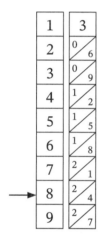
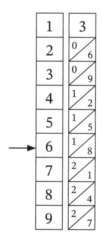

3. $63 ÷ 7 = $9. You can spend $9 a person.

188   REVEALING ARITHMETIC

MULTIPLICATION: MULTI-DIGIT OPERATIONS

Multiplication in Real Life

1. $9 \times 20 = 180$. You can go 180 miles.

2. $9 \times \$50 = \$450$. It will cost you $450.

3. $148 \times \$3 = \$444$. It would cost the church $444.

4. $52 \times \$5 = \$260$. You would have $260, assuming you set aside $5 every week. (There are 52 weeks in a year.)

Apartment Rental

	Furnished	Unfurnished with Rented Furniture	Unfurnished with Delivered Furniture
Price for First Month	$ 2,600	$ 2,000	$ 3,400
Recurring Price After First Month	$ 2, 600	$ 1,900	$ 800
Price for 4 Months	$10,400	$ 7,700	$ 5,800
Price for 7 Months	$18,200	$13,400	$ 8,200
Price for 12 Months	$31,200	$22,900	$12,200

1. Which option would be the most economical if you end up staying 1 month? **Rented Furniture**

2. What is the difference between a furnished apartment and an unfurnished apartment with rented furniture for 1 month? **$600**

3. What is the most economical option for 7 months? How much more expensive are the other options? **Unfurnished with delivered furniture is the most economical option for 7 months. An unfurnished apartment with rented furniture would be $5,200 more, and getting a furnished apartment would be $10,000 more.**

DIVISION: MULTI-DIGIT OPERATIONS

Division in Everyday Life (Version A)

1. a. Each book cost $4.
   ($200 cost ÷ 50 books = $4 cost per book)

   b. They would charge $32.
   (8 × $4 cost = $32)

2. You should read 20 pages each day.
   (80 pages ÷ 4 days = 20 pages)

3. You went 10 miles per gallon.
   (90 miles ÷ 9 gallons = 10 miles per gallon)

Division in Everyday Life (Version B)

1. a. Each book cost $0.80.
      ($16,000 cost ÷ 20,000 books = $0.80 cost per book)

   b. They would charge $6.40.
      (8 × $0.80 cost = $6.40)

2. You should read 7.5 pages each day.
   (150 pages ÷ 20 days = 7.5 pages)

3. You went 24.75 miles per gallon.
   (297 miles ÷ 12 gallons = 24.75 miles per gallon)

DECIMALS

Trip Planning, Part 1

Time of Trip if Average 55 MPH rounded to the nearest hour:
approximately 31 hours (1,712 ÷ 55 ≈ 31)
Time of Trip if Average 65 MPH rounded to the nearest hour:
approximately 26 hours (1,712 ÷ 65 ≈ 26)

Total Gas:	$258.86	(86 × $3.01)
Total Hotel:	$91.50	($85 + $5 + $1.50)
Total Food:	$47	($0 + $10 + $15 + $4 + $8 + $10)
Total Other Expenses:	$12	($4 + $8)
Cost of Trip:	$409.36	($258.86 + $91.50 + $47 + $12)

Trip Planning, Part 2

Answers will vary. You can check your child's answers by adding up the numbers in each individual section, then checking to make sure they equal the total. See Part 1 for an example.

Math and Our Bodies

Note: Student was instructed to round answers to the nearest hundredth.

1. 2.41 times   (60,000 miles of blood vessels ÷ 24,900 miles = 2.41)

2. 23.22 times   (60,000 miles of blood vessels ÷ 2,583.9 miles = 23.22)

RATIOS & PROPORTIONS

The Way Things Grow

1. The man's head is approximately 2 centimeters, and the baby's head is approximately 1 centimeter.

2. The distance from the man's neck to the bottom of his shoe that is not stepping forward is approximately 11 centimeters; the distance from the baby's neck to the bottom of his foot is approximately 3 centimeters.

3. Ratio between man's head and his body: 2:11 or $\frac{2}{11}$

   Ratio between baby's head and his body: 1:3 or $\frac{1}{3}$

Notice that the two ratios are not equal. The baby's head is much larger in comparison to his body than the man's.

Math and Sunflowers

Note: Student was instructed to round answers to the nearest hundredth.

Sunflower A: 1.63 ($\frac{13}{8} \approx 1.63$)
Sunflower B: 1.62 ($\frac{34}{21} \approx 1.62$)
Sunflower C: 1.62 ($\frac{55}{34} \approx 1.62$)
Sunflower D: 1.62 ($\frac{89}{55} \approx 1.62$)
Sunflower E: 1.62 ($\frac{144}{89} \approx 1.62$)

EXPONENTS & SQUARE ROOTS

Note: Student was told to round the answer to the nearest hundredth.

$a^2 + b^2 = c^2$	1. Pythagorean theorem
$5^2 + 3^2 = D^2$	2. Substituted values.
$34 = D^2$	3. Simplified.
$D = \sqrt{34}$, approximately 5.83	4. Took square root of both sides.

The sailor will need to travel approximately 5.83 miles if he takes a diagonal line.

Appendix A:
Mathematicians

Looking at famous mathematicians and studying their worldviews and mathematical accomplishments serves multiple purposes. It shows the importance of what we believe, stimulates thinking of math outside a textbook, and provides an opportunity to recognize and sift through different worldviews. This appendix is designed to give you a few pointers and ideas to get you started.

When looking for information on a mathematician, try the Internet, an encyclopedia, or your local library. I have listed a few resources I found helpful on the Book Extras page of ChristianPerspective.net.

SHORT LIST OF MATHEMATICIANS

Unsure which mathematician to study? Here is a short list.

- Archimedes
- Aristotle
- Charles Baggage
- René Descartes (comparing and grouping numbers — footnote on page 38)
- Albert Einstein (subtraction — page 50)
- Euclid
- Leonhard Euler
- Carl Gauss
- Gottfried Wilhelm Leibniz
- Immanuel Kant
- Johannes Kepler (fractions — page 113)
- John Napier (multiplication — pages 64 through 69)
- Sir Isaac Newton (fractions — page 113)
- Blaise Pascal
- Leonardo "Fibonacci" Pisano (Appendix D — page 212)
- Plato
- Pythagoras (exponents — page 162)

QUESTIONS TO ASK

Here are a few questions you could ask yourself or have your child ask himself while studying a mathematician as a springboard for discovering lessons from his life and work.

- ◆ What mathematical accomplishments did this man make? How and why did he make them? How do we use his contributions today?

- ◆ How did this man view math? How did this man's view affect how he used math? Did he use his accomplishments practically? Why or why not?

- ◆ How did this man's worldview line up with scripture? What were some truths he believed? Some lies?

- ◆ How do we see the hand of God at work in this man's life?

- ◆ What are some commendable character traits or beliefs we could learn from this man? What were this man's negative character traits or beliefs we should be careful about in our own lives and thinking?

- ◆ What are at least three lessons we could learn from either the positive or negative in this man's life?

Appendix B:
Different Number Systems

Over the years, men have used many creative and different techniques to record quantities!

Below are brief explanations of just a few of the many different number systems that have been or are used. I have included more details on ones I thought might be fun to explore. Even more details on many of them can be found by searching the Internet.

Number systems below are arranged in two categories (fixed value systems and place value systems). Within each category, I have listed the systems alphabetically. In an effort to keep the explanations simple (and usable), I have not covered all the different variations of the systems.

FIXED VALUE SYSTEMS

In a fixed value system, each symbol has a fixed value that does not change based on where you place it. The Aztec, Egyptian, and Hebrew systems are three examples.

Aztec Indian

The Aztec Indians used a dot to represent 1, a diamond for 10, a flag or hatchet[98] for 20, portions of a feather for the hundreds, and a purse/bag with tassels for 8,000. The Aztecs placed their larger numbers to the left of the smaller numbers, just as we do today.[99] And notice the use of a vertical line to separate the one symbols into a set of 5 and an additional 1, which makes it easier to quickly count up. It is unclear in the resource I consulted[100] if this was always done, but an example is given where it was.

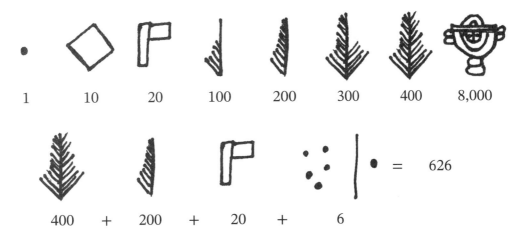

---

98. Lucien Biart, *The Aztecs*, page 319 (quoted in Cajori, *History of Mathematical Notations*, 1:41), refers to it as a flag, while Jan Gullberg, *Mathematics: From the Birth of Numbers*, page 60, a hatchet.
99. Images and information based on that given in Cajori, *History of Mathematical Notations*, 1:41, and in her quote from Lucien Biart, *The Aztecs*, trans. J.L. Garner (Chicago, 1905), 319.
100 Cajori, *History of Mathematical Notations*, 1:41.

Egyptian

The chart below shows some Egyptian numerals. Not everyone, however, always used the same numerals or drew the numerals the same way; the Egyptians also had other variations and writing styles.[101]

$$0 = 1 \qquad \text{(symbol)} = 100$$
$$\text{(symbol)} = 10 \qquad \text{(symbol)} = 1{,}000$$

Notice how the symbol for 1 looks like a vertical staff, the symbol for 10 looks like a heel bone, the symbol for 100 looks like a rope coil or a scroll, and the symbol for 1,000 resembles a lotus flower.[102] Notice the use of common symbols to represent numbers.

To write numbers lacking their own symbol (like 2, 7, 30, etc.), the Egyptians repeated other symbols. The Egyptians always placed the smaller values on the left and the larger ones on the right. To read the number correctly, you need to read from right to left. Notice in the examples below that the symbol for ten (𓎆) is to the right of the symbol for one (𓏤).

22	𓏤𓏤 𓎆𓎆
1,491	𓏤 𓎆𓎆𓎆𓎆𓎆𓎆𓎆𓎆𓎆 𓏢𓏢𓏢𓏢 𓆼

Hebrew

The Hebrew number system is too complicated to go into depth here, but I could not fail to mention it briefly since Hebrew is the language in which the Old Testament was first written. In Hebrew, the alphabet doubles as a number system. Thus the Hebrew letter for *a* also means *1*, their letter for *b* also means *2*, etc. There are enough letters for the digits 1-10; for 10, 20, 30, and so on up to 100; and, with some variations on the letters, for the hundreds. Quantities are formed by combining these letters, as shown. Hebrew numerals are read from right to left.[103]

$$הל = 35$$
$$ל = 30 \qquad ה = 5$$

---

101. The hieroglyphic style shown here was the oldest, more decorative style, and the one typically associated with the Egyptians. However, the Egyptians also developed two cursive styles (hieratic and the demotic) that were faster to write. Notation within the styles also varied. See Cajori, *History of Mathematical Notations*, 1:11, and Smith, *History of Mathematics*, 2:47.
102. These resemblances are those pointed out in Groza, *Survey of Mathematics*, 38.
103. Cajori, *History of Mathematical Notations*, 1:19–21.

PLACE VALUE SYSTEMS

In a place value system, the place, or location, of a symbol determines its value. The Hindu-Arabic decimal system with which most of us are familiar is a place value system. Below are a few other different place value systems.

Babylonian

The Babylonians used ⟨symbol⟩ for 1 and ⟨symbol⟩ for 10. They combined these symbols to make other numbers.

The problem the Babylonians faced was that for many years at least they did not use any symbol to represent zero. This left a lot of interpretation up to context. ⟨symbol⟩ could have meant either 12 or 602 [(10 × 60) + 2]. Likewise, the example shown below could have meant either 52 or 3,002 [(50 × 60) + 2].

Binary and Hexadecimal

Did you realize computers operate off a place value system called the binary system? This system uses just two symbols — 1 and 0 — to represent every number. Because of the way computer circuitry is hard wired, these symbols translate to the computer as "on" and "off" flows of electricity.

Although binary numbers translate well to electrical pulses, they tend to get long quickly, making them impractical for us to read. To help make numbers more readable, computer programers often use hexadecimal numbers (a system based on 16) to represent binary numbers. For example, 255 is written 11111111 in binary, but can be simplified by someone familiar with the systems to the hexadecimal number "ff."

If you have any interest in the workings of computers, you may want to do an Internet search for these two number systems. My brother, who loves computers, had hours of fun learning these systems.

Incan

Instead of developing a system to write numbers, the Incas recorded quantities by tying knots on a device called a quipu (kē′pōō).[104] The quipu system was extremely complicated, and only special quipu makers, called quipucamayocs, were able to interpret them.

---

104. Pronunciation from *The American Heritage Dictionary of the English Language*, 1980 New College Edition, s.v. "quipu."

Although we do not know a lot about quipus, we do know they used place value. Quipu makers would represent the number 36 by tying three knots near the top of the chord, and six knots near the bottom of the chord. They used different colors and sizes to represent different types of records. The location of the chord and type knot also affected the number represented.[105]

Apparently, the Incas were very successful with this innovative approach to record keeping. They operated a huge empire that included more than 15,000 miles. For instructions on making a simple quipu, see page 34.

Roman

In the Roman numeral system, a single quantity (one) is represented by a capital I. To represent two quantities, repeat the symbol twice (II). However, because it would not be easy to read large numbers this way (imagine trying to read IIIIIIIIIIIIIIIIIIIIIIIIIIII!), other symbols are used to help represent larger numbers. The picture shows different symbols in the Roman numeral system.

I	1
V	5
X	10
L	50
C	100
D	500
M	1,000

To represent a quantity that does not have a specific symbol (like three) using this system, you would combine several symbols. Three is written III (1 + 1 + 1).

When writing in Roman numerals, it is customary not to repeat the same symbol more than three times. Instead of writing four as IIII, we would put a symbol for one (I) in front of the symbol for five (V), like this: IV. Whenever a smaller value is to the *left* of a larger value, it means the smaller value should be subtracted from the larger value.

$$I = 1 \qquad V = 5 \qquad IV = 5 - 1 = 4$$

When the smaller value is to the *right* of the larger value, it means the smaller value should be added to the larger value.

$$VI = 5 + 1 = 6$$

This is a custom adopted to make the system standard and easier to read and write. It is much easier to read and write IX than it is to write VIIII, as there are fewer symbols involved.

Interesting Note: Roman numerals did not always appear in their present form. There was a time when "four" was written IIII instead of IV.

---

105 Marcia and Robert Ascher's *Mathematics of the Incas*, 31, 33.

Appendix C:
Abacuses

Abacuses are wonderful manipulatives to visually reinforce mathematical concepts. They help children connect paper symbols with actual quantities. I still remember the fun my mom and I had making abacuses when I was little — I had no idea then how much I was learning! Take a brief look with me at what an abacus is and what it does, after which we will discuss how to make one or find one to use.

According to the *Merriam-Webster Online Dictionary*, an abacus is "an instrument for performing calculations by sliding counters along rods or in grooves."[106] The actual makeup of abacuses vary greatly. There are many types of abacuses, including counting boards and bead abacuses. Within these broad types, there are specific variations, such as the Roman, Chinese, and Japanese.

But although variations abound, the basic principles remain the same. Some sort of marker (bead, pebble, etc.) is moved along either a rod, wire, line, or groove. Each separate rod/wire/line/groove in the abacus (and sometimes separate sections within those rods/wires/lines/grooves) has a different place value.

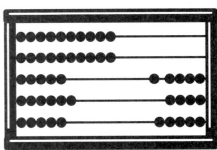

"545" on a home-made abacus (See explanation under "Using your Abacus.")

Once proficient with their use, abacuses can save significant time. Years ago, merchants would use abacuses to quickly add totals — in fact, our term "checkout counter" may have originated to describe the abacus counting table used to add up a buyer's totals.

You can easily make a simple abacus using either an old picture frame (see instructions following this introduction). These homemade abacuses are wonderful for teaching younger children how to count by groups, understand place value, and perform simple calculations. I also think young children — especially hands-on learners — benefit from making an abacus themselves. The process of stringing an abacus can be used to reinforce place value, counting skills, and grouping skills.

Japanese or Chinese abacuses (also called Japanese sorobans and the Chinese suan pans), while not easy to make, are very affordable to purchase and excellent tools. You will notice these abacuses have a horizontal divider. Each bead above the divider represents five of the unit (five ones, five tens, five thousands, etc.), while each bead below the divider represents one of the unit. By being able to quickly represent groups of five, these abacuses speed up operations significantly and make these abacuses easier to use as actual computational instruments instead of only learning tools. They also help students

---

106. *Merriam-Webster Online Dictionary*, s.v. "abacus" (2009), http://www.merriam-webster.com/dictionary/abacus (accessed February 22, 2009).

learn to perform mental arithmetic faster by training them to think in terms of how numbers add to get to ten. If you search the Internet for "abacus buy," you should be able to find several companies selling this type of abacus. Because they require thinking in terms of groups of five instead of just groups of 10, which is different than how most of us in the Western world learned math, they are not as intuitive to pick up. Because of that, we won't be covering them here. If you choose to use one with your child, you should be able to find instructions online as to how to use them.

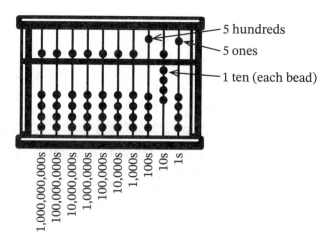

"545" on a Japanese/Chinese Style Abacus

Another option is to make a counting table abacus. These abacuses, which were used extensively in Western Europe at one time, consisted of tables with either horizontal or vertical lines upon which "counters" could be moved to perform calculations. Instructions are included later in this appendix for making a simple counting table abacus out of poster board.

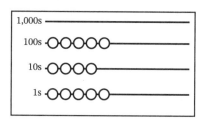

"545" on a counting table abacus

In addition to making or buying an abacus, your child may enjoy playing around with online abacuses. There are many applets online that allow users to customize the number system, size, and type of abacus used. See the Book Extras page on ChristianPerspective.net for some, or search online.

MAKING AN ABACUS

WARNING: These abacuses contain small parts (beads) that can be a choking hazard as well as wires/nails that could hurt if handled inappropriately; please be careful when using around young children.

Picture-Frame Abacus

Supplies:

- **Wooden frame** — You can use a 5 × 7 or larger picture frame with the glass removed, or make your own frame out of 1 × 2s. For a medium-sized abacus, try an 8 × 10 frame.

- **Multi-color beads** — Basic pony beads will work — look in the craft section of your local department or craft store. The number of beads you need depends on the size of your frame. You need 50 beads for an 8 × 10 frame.

- **Wire** — You can use plant wire, stripped electrical wire, or any sort of thin, flexible wire you can wrap around a push pin. Make sure the wire is thin enough to easily twist around the push pin. Alternately, if you have a thick enough picture frame to drill holes into, you can use any sort of thick wire that is sturdy enough to insert into drilled holes. It is much harder to find a frame thick enough to drill into, but drilled frames will make it easier to restring the abacus different ways to teach different skills.

- **Needle-nose pliers and push pins**, or, if using thicker wire, a drill

Instructions:

1. Decide how many rows you want and cut the wire into strips a few inches longer than the width of your frame. Five is a good number of rows for most medium frames; really large frames can handle more.

2. Mark the frame at evenly spaced intervals along both sides where you want your rows to be. (Here is where those other math skills come in handy.)

3. Prepare the frame for the wire by either inserting push pins at each of the marks, or else drilling holes in the frame. A lot will depend on what type of frame and wire you have. You must have a sturdy frame and wire to drill holes; otherwise, you will need to use the push pins.

4. Secure one end of the wire by wrapping it around the push pins, or by pushing a thicker wire into the drilled holes.

5. Add the beads to the first row of the abacus. You can arrange the beads in various ways depending on what you are trying to teach (see the "Fun Variations" section below). For a basic, all-purpose abacus, it is nice to alternate between 5 beads of one color and 5 beads of another color. For a typical abacus, put 10 beads per row.

6. Secure the second end of the wire to the frame the same way you did in step 4.

7. Repeat steps 4-6 until you have completed all the rows.

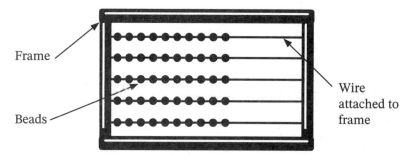

Each bead on the first wire of the abacus represents one unit; in the second, ten; in the third, one hundred; and in the fourth, one thousand. To represent a number, simply move the appropriate number of beads or objects to one side of the abacus. The picture shows how you would represent "545" by moving 5 beads from the hundred's row, 4 from the ten's row, and 5 from the one's row. See the "Using Your Abacus" section for information on adding/subtracting with an abacus.

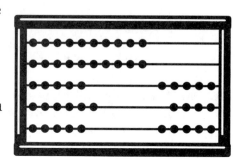

Fun Variations

There are a lot of fun variations you can do with an abacus. Feel free to experiment with larger/smaller abacuses — and with different colored beads. Color coding the beads on the abacus can further help your child grasp concepts.

For example, if you are trying to teach your child to count by groups (2s, 3s, 4s, etc.), have him string an abacus using different colored beads. You might have him string one row with alternating groups of two, the next with alternating groups of three, and so forth. Adjust the number of beads on each row as necessary to form even groups.

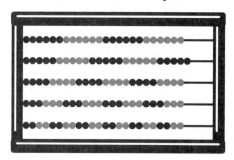

You could also color code an abacus to help with decimal numbers. Draw a little decimal point on the frame. Use all one color bead to represent numbers less than one, and a different color to represent numbers greater than one.

202 REVEALING ARITHMETIC

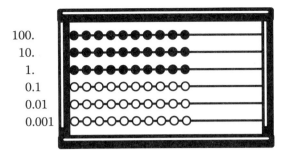

Be creative and have fun! Although it is often helpful to make more than one abacus, you can reuse the same abacus over and over again by restringing it different ways.

Poster Board Counting Table Abacus

Supplies: Poster board, black marker, objects (rocks, LEGOs, dried beans, or any small object)

Instructions: Draw four lines on a piece of poster board and label them as shown. Use objects (rocks, LEGOs, dried beans, or any small object) as markers to represent quantities. The picture shows how you would represent 545. See the "Using Your Abacus" section below for information on adding/subtracting.

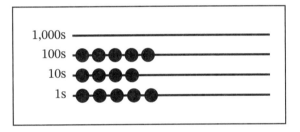

Fun Variations

Since these abacuses were commonly used with Roman numerals, the spaces in between the lines were used to represent 5s, 50s, etc. If you like, you could label your abacus that way as well (shown below on the top left) — or even make one labeled with Roman numerals (top right). You can also vary the orientation of the lines — some counting tables had vertical lines instead of horizontal ones (bottom left). Others had vertical lines in addition to the horizontal lines, leaving multiple workspaces (bottom right).[107] Feel free to have fun and develop your own variations! If you like woodworking, you could make a counting table abacus out of a wood board (smoothed to avoid splinters) instead of poster board.

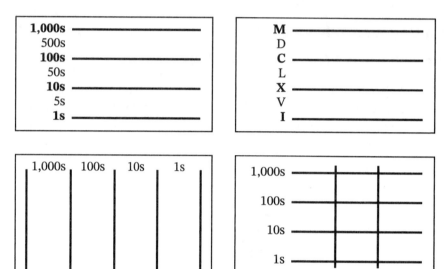

---

107. On these abacuses, to add 20 + 4, the 20 could be formed in the first workspace, and the 4 in the second. The third workspace would be used to form the addition of the two.

USING YOUR ABACUS

This section contains instructions on using an abacus. For more details or information on how to use a Chinese/Japanese style abacus, search the Internet for "abacus how to." Numerous resources, including online videos, are available, especially for the Chinese/Japanese style abacus (which we did not show here since their use requires approaching problems differently than with our typical addition/subtraction/multiplication/division methods, making them initially more challenging for those used to the typical methods to learn).

ADDITION
Solving 545 + 116 = 661

Note: White beads show the beads moved.

Picture Frame Abacus   **Poster Board Abacus**

Step 1: Form the starting quantity, in this case, 545.

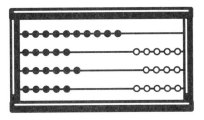 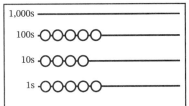

Step 2: Add the second quantity, in this case, 116.

While adding the 6 beads on the one's row, we run out of beads after adding 5.

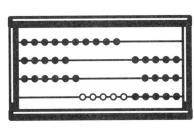 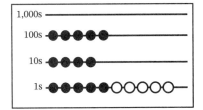

So we exchange 10 beads on the one's row for 1 bead on the ten's row.

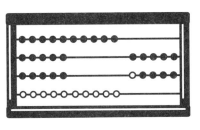 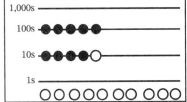

Now we can add the remaining 1 bead on the one's row, as well as a bead on the ten's and hundred's row for the 1 ten and 1 hundred in 116.

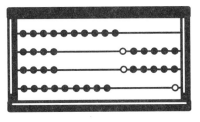 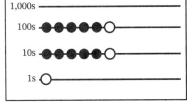

Note: You may also see abacuses solved starting at the greatest place value (so the hundred's here) down to the least. We chose to show it this way as that's how we solve using the traditional method on paper.

APPENDIX C: ABACUSES

SUBTRACTION
Solving 545 – 116 = 429
Note: White beads show the beads moved.

Picture Frame Abacus   **Poster Board Abacus**

Step 1: Form the starting quantity, in this case, 545.

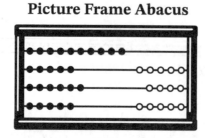 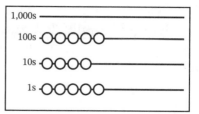

Step 2: Subtract the second quantity, in this case, 116, borrowing as necessary.

We want to subtract the 6 ones in 116. We only have 5 beads to the right on the one's row, so we'll start by subtracting them.

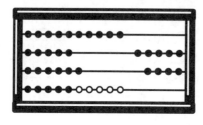 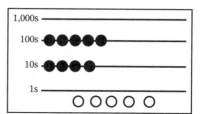

We still have 1 more to subtract from the one's row, as we need to subtract 6 and have only subtracted 5. So we'll exchange 1 bead on the second row (the ten's row) for 10 beads on the first row (the one's row).

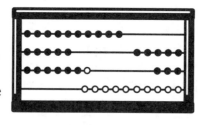 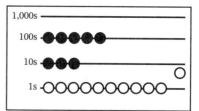

Now we can subtract the remaining 1 bead from the one's row.

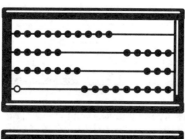 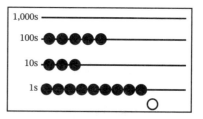

Next, we need to subtract the 1 ten in 116 and the 1 hundred in 116 by moving/ taking away 1 bead from the ten's row and 1 bead from the hundred's row.

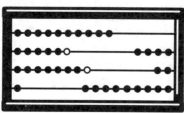 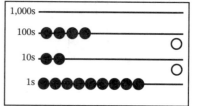

206   *REVEALING ARITHMETIC*

MULTIPLICATION
Solving 24 × 3 = 72
Note: White beads show the beads moved.

You can multiply on an abacus by remembering that multiplication is just repeated addition. You want to always use the smaller number as the multiplier, so we'll look at 24 × 3 as 3 × 24, which means 3 sets of 24, or 24 + 24 + 24. (The order doesn't matter in multiplication, so that will equal the same thing as 24 sets of 3.) To complete the multiplication, just add 24 three times! You can use the top row of your abacus to keep track of how many times you've added 24; just move a bead from the top row to the right each time you add 24 (or start with 3 beads on the top row, and move one back each time you add 24 until you have none left).

Picture Frame Abacus   **Poster Board Abacus**

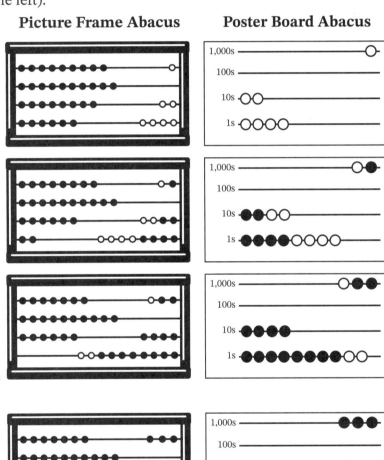

Step 1: Form the greater amount (24 in this case) on the abacus and move/put 1 bead from the top row to show we have 1 set of that quantity.

Step 2: Add that same quantity again and move/put 1 bead on the top row to show we have 2 sets of the quantity.

Step 3: Add the quantity again and move/put 1 bead on the top row to show we have 3 sets of that quantity. Note that we only have 2 beads left on our one's row, so we will have to add those 2.

We'll then exchange 10 beads on the one's row for 1 on the ten's row.

We can then add the remaining 2 one's, as well as the 2 ten's in 24. Now that we've formed 3 sets of 24, we have only to read the number formed (72) to see the answer!

APPENDIX C: ABACUSES 207

DIVISION

Solving 16 ÷ 3 = 5r1 or $5\frac{1}{5}$

Note: White beads show the beads moved.

You can multiply on an abacus by viewing division as repeated subtraction. Just subtract the amount you're dividing by repeatedly until there is none left, using the top row of the abacus to keep track of how many times you had to subtract.

Picture Frame Abacus **Poster Board Abacus**

Step 1: Form the amount you're dividing on the abacus (16 in this case).

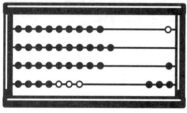 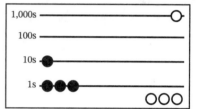

Step 2: Subtract the amount you're dividing by (3 in this case), moving/putting a bead on the top row to show that you've subtracted it once.

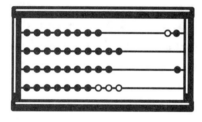 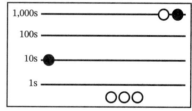

Step 3: Continue subtracting the amount you're dividing by (3 in this case), moving/putting a bead on the top row each time until you can't subtract it anymore.

Note that we run out of beads on the one's row, so we have to exchange 1 bead from the ten's row for 10 beads on the one's row.

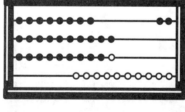 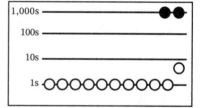

Now we can continue to subtract 3 more.

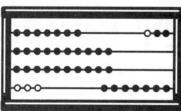 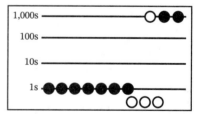

208 REVEALING ARITHMETIC

We subtract 3 more again.

We subtract 3 more again.

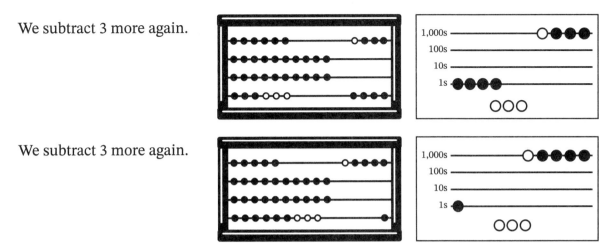

We can't continue dividing, as we can't subtract any more sets of 3. We see that the answer is 5 with 1 remaining, as the top row of the abacus has 5 beads on the right, telling us that we subtracted 5 times, and we have 1 bead still remaining on our one's column. We essentially formed 5 sets of 3, and ended up with 1 left over or 1 we still need to divide by 5 (which we could write as $\frac{1}{5}$ — the fraction line can be thought of as meaning "divided by").

Appendix D:
Math Methods

The more I have looked at different math methods (also called algorithms), the more amazed I have become. There are so many different ways to solve addition, subtraction, multiplication, and division problems!

Sadly, many students look upon the math methods taught in their math book as *the* way to solve math problems. I know for years I did.

Although I did not realize it at the time, viewing math methods as *the* way to solve math problems was subtly reinforcing a very wrong and dangerous view of math itself. Because I looked upon math methods as absolutes in themselves (as opposed to ways of describing God's absolutes) and as something man created and enabled to work (as opposed to a method man thought of to help him record the principles God holds in place), I unknowingly let math encourage me to trust numbers and man instead of God. I completely missed seeing God in math.

In this appendix, I would like to explore a few math methods in more depth. My goal in doing this is to provide you with information you can use to show your child that the method in his textbook is not some sort of magical system that makes math work, but rather just one method out of many we can use to record the consistencies God holds in place.

Looking at various math methods also has other side benefits. Learning different ways to solve problems can help stimulate creativity and teach students to really think mathematically rather than memorize a rule. Historical methods may also help interest non-mathematical students in math. A child struggling with the math method in his textbook may even find a method he finds easier to use.

That said, I should also include a note of caution. Looking at alternate math methods may confuse a child just learning a math operation. Please use your own judgment as to when to introduce these other methods to your child — and as to how much depth to cover.

This appendix is by no means exhaustive — it is merely a sampling. My goal here is to show that our modern math method is *one* way of describing the consistencies God has put around us. Please see Appendix B for information on other ways to write numbers — I have focused here only on methods for solving equations.

Note: From here out, this appendix is written as if to the student. I have done this to make it easier to read and explain, and also to make it useful for older students.

ADDITION AND SUBTRACTION METHODS

Let us say we need to add 12 and 19 or that we need to subtract 14 from 21. There are a lot of different ways we could do this! Below are a few methods we could use — we could also use non-written methods like an abacus or calculator.

Methods Covered:
- A. Traditional
- B. *Liber Abaci*
- C. *Speed Mathematics Simplified*
- D. Bhāskara Method

A. Traditional Method

The figures show how most modern textbooks present addition and subtraction. The numbers being added or subtracted are written on top of each other, and the problems are solved from right to left. In addition, we "carry" any groups of 10 we get while solving the one's column over to the ten's column; in subtraction, we "borrow" groups of 10 from the ten's column as needed. The 2 in 21 is crossed out and replaced by a 1 because one group of ten was taken over to the one's column to form 11. See **ADDITION AND SUBTRACTION: MULTI-DIGIT OPERATIONS** for a more detailed explanation.

$$\begin{array}{r} \overset{1}{1}2 \\ +19 \\ \hline 31 \end{array} \qquad \begin{array}{r} \overset{1\ 1}{\cancel{2}1} \\ -14 \\ \hline 7 \end{array}$$

B. *Liber Abaci* Method

Leonardo Pisano, nicknamed "Fibonacci," was born in Italy in 1170, right in the middle of the Medieval Ages. While Leonardo was still a young boy, his father's job took the family across the Mediterranean Sea to Bugia (a port in northern Africa) where Leonardo learned a new way of computation. In most of Europe, people performed computations using abacuses, but in Bugia, Leonardo was exposed to the Hindu method for computation using the digits 0-9 and written methods, or algorithms. After becoming a proficient mathematician and studying math in great depth, Leonardo wrote a book titled *Liber Abaci* (1202) with the goal of sharing the Hindu system with his native Italy.

This famous book greatly helped bring the number system we use today not only to Italy, but through Italian merchants, to the rest of Europe, paving the way for men like Galileo Galilei and Isaac Newton to make their famous discoveries.

Within *Liber Abaci,* Leonardo presented a method for addition using both writing and a finger system almost like sign language. He would first write the two numbers to be added on top of each other, then solve the problem from right to left, bottom to top, writing the sum *above* the two numbers being added.

To solve 19 + 12 using this method, first write 12 on top of the 19, then add 9 + 2. Since 9 + 2 equals 11, write a 1 above the one's column and hold up your left hand index finger to remind you to carry the 1 (which represents the 10 in 11; hence, it is representing 1 group of 10 and belongs in the ten's column) to the ten's column.[108] Next, add the ten's column (1 + 1), remembering to add the 1 brought over from 11. Write the answer to the ten's column (3) above the two numbers being added to form the final answer, 31. Note: The word "Sum" in the figure is added just for ease of reading.

```
31 ←— Sum
12
19
```

Leonardo's subtraction method was very similar to his addition method. He wrote the numbers to subtract, then wrote the answer *above* them. If he needed to do any borrowing, he would use his left hand to handle borrowing, although a bit differently than we do today.

To solve 21 - 14 = 7 using his method, write the two numbers to be subtracted on top of each other, then work the one's column. Since 1 is less than 4, mentally move 10 from the ten's column to make 11 - 4. Write the answer, 7, above the line. To remember you took from the ten's column, hold up your left hand index finger[109] (representing 1 group of 10).

```
 7
---
21
14
```

Lastly, work the ten's column. Since you had to take ten from the ten's column, *add* ten to the number being subtracted (in this case, the 1 in 14). This gives 2 - 2, or 0. Since the answer is zero, you do not have to write anything. Note: Adding to the bottom number does the same thing to the answer as subtracting from the top number would. It is just a different way to think about it.

If you would like to practice solving problems using Leonardo's methods, you can do so using sign language to "carry" and "borrow" numbers in your left hand. An online search for "sign language numbers" should provide several pictures you can use as a reference to know how to form the numbers. While American Sign Language is not quite the finger system Leonardo used, it should work quite well.

Leonardo's methods looked quite foreign to the people who first read *Liber Abaci*. People of the time were used to using abacuses to solve everything, not written math methods. But gradually, Europe switched to written math.

Whether Leonardo acknowledged it or not, God had his hand over his life and used him to accomplish His purposes. Although Leonardo may not have understood why as a little boy he had to leave his hometown and go to Africa, God used his father's job to expose the young boy to a new system of numbers. He then used Leonardo's work to help bring the Hindu-Arabic number system to Italy, and from there, to Europe.

---

108. This is the way to represent "1" in American Sign Language. In Leonardo's actual system, "1" would have been "carried" by bending the left hand small finger.
109. See previous footnote.

APPENDIX D: MATH METHODS

C. *Speed Mathematics Simplified* Method

In his *Speed Mathematics Simplified*, Edward Stoddard presents a speed-solving method based on the steps followed when solving problems using a Japanese abacus (also called a soroban — see Appendix C). This method works from left to right rather than right to left, and takes advantage of complements (numbers that add together to equal 10) to make adding numbers easier.

Since this method is quite different, it may take a bit to become proficient. But once mastered, it will let you solve equations much easier and faster. If you decide to learn the method, I would suggest purchasing a Japanese soroban (see Appendix C) and reading *Speed Mathematics Simplified*.

To add using this method, write the numbers being added and draw a line below them much as you would in the traditional method. But instead of working from right to left, work the problem from *left to right*. First add the ten's column, and write down the answer (2). Now move to the right, where you encounter 2 + 9. Instead of thinking "2 + 9 equals 11," think about it in terms of what makes 10. You know right away that 1 + 9 makes 10, so you know you have 1 group of 10 with 1 left over. Draw a tiny mark underneath the 9 to represent the group of 10, and write a 1 underneath the line to represent the 1 you have left over.

```
 12
 19
 21
  1
 31
```

Now add up all the tiny marks you have (each mark represents a group of 10) and write that number (in this case, 1) in the ten's column. Add the numbers below the line to get your answer.

I know this might not seem faster than the traditional method, but it truly is once you get used to it — especially when adding lots of numbers!

To subtract using the speed method, first write the numbers being subtracted, then solve, again from *left to right*. In this case, we would start in the ten's column. Here we have 2 groups of 10 minus 1 group of 10, which equals 1 group of 10, so we would write a 1 underneath the ten's column to represent 1 group of 10. When we get to the one's column, we see we need to minus a larger number (4) from a smaller number (1), so we cross out the 1 from the answer we wrote in our ten's column. But rather than adding that 10 to our one's column and thinking "11 − 4 = 7", we automatically subtract the 4 from the 10 we scratched out by taking the complement (number that added to it would equal 10) of 4, which is 6, and adding it to the 1. We then have our answer: 7.

```
 21
 14
 ──
 ₁7
```

D. Bhāskara Method

This method is attributed to Bhāskara II, a mathematician from India.[110] The figure shows how to add 12 + 19 using this method. First, the one's column is added (9 + 2 = 11), then the ten's

Bhāskara Method:
$$2 + 9 = 11$$
$$1 + 1 = \underline{20}$$
$$31$$

Traditional Method (for comparison):
```
  ¹
  12
 +19
 ───
  31
```

---

110. Groza, *Survey of Mathematics*, 215. The footnote adds that, "The Hindus used the dot, ·, as the symbol for zero." A zero was used here instead of a dot for simplicity.

column is added (1 + 1 = 2). Lastly, the two columns are added together. Instead of writing the problem vertically, it is written horizontally. The traditional method is shown for comparison.

MULTIPLICATION METHODS

Below are a few of the many multiplication methods out there. See page 64 for an explanation of Napier's rods and the multiplication table.

The methods included here are truly just a sampling. I hope you have fun exploring them... and perhaps even thinking up your own method!

Methods Covered:

A. Traditional
B. Dust Board
C. *Liber Abaci*
D. Chessboard

E. Gelosia
F. Duplation Method
G. *Speed Mathematics Simplified*
H. Juan Diez Method

A. Traditional Method

The traditional method is simply a "rule" to help us keep track of digits as we multiply. When we multiply 12 × 9, we are really finding what 9 groups of 2 equal (9 × 2) and what 9 groups of 10 equal (9 × 10) and adding the answers together to find 9 × 12. The algorithm, or method, we use helps us do this quickly without losing track of the value of each number.

```
  1
 12
× 9
───
108
```

We write the two numbers on top of each other, then multiply the digits from right to left. 9 × 2 equals 18, so we write an 8 underneath and a 1 over the ten's column. (We put the 1 above the ten's column because it represents 1 group of 10!) We then multiply 9 × 1 (the 1 represents 1 group of 10), and add the answer (9 groups of 10) to the 1 group of 10 we carried over from the previous column, giving us 10 (groups of 10), which we write below in the ten's column next to the 8 to make 108.

B. Dust Board Method

The Hindu mathematicians did not have lots of paper at their disposal like we do today. They performed their calculations on a dust board. Hence, erasing each step as they went made a lot of sense. You can see this quite clearly in this method, which comes from *Principles of Hindu Reckoning: A Translation with Introduction and Notes by Martin Levey and Marvin Petruck of the Kitāb Fī Usūl Hisāb Al-Hind.*[111]

» Step 1: Write the two numbers to be multiplied on top of each other.

```
12
 9
```

---
111. Kūshyār ibn Labbān, *Principles of Hindu Reckoning*, 14–16.

» Step 2: Multiply 9 × 1, erasing the 1 and writing the answer (9) in its place.

$$92$$
$$9$$

» Step 3: Erase the bottom 9 and move it underneath the 2. Moving the number to the right helps us remember which numbers we have finished multiplying (very important, especially when working with numbers with more digits).

$$92$$
$$9$$

» Step 4: Multiply 9 × 2, mentally adding this to 90 (the answer from Step 2 — the 9 we wrote was in the 10's place) and writing the final answer, 108.

$$108$$
$$9$$

C. *Liber Abaci* Method

Note: See page 212 for an overview of *Liber Abaci*.

The multiplication method presented in Leondardo Pisano's *Liber Abaci* involved a combination of writing and forming numbers using a sign language type system with the left hand. To multiply numbers using this method, first write the numbers to be multiplied, putting the greater number on the *bottom*. Then work right to left, using your left hand to "carry" any numbers from one column to the next and writing the answer on top of the two numbers being multiplied.

For example, to multiply 12 × 9, first write the numbers down. Then multiply 9 × 2, writing an 8 on top and forming a 1 using the left hand by holding up your index finger[112] (representing the 10 in 18). Next multiply 9 × 1, adding to it the "1" carried on the left hand for a total of 10, which is written next to the 8 to make 108.

$$108$$
$$9$$
$$12$$

Note: If you would like to solve problems using this method, you can use American Sign Language to "form" the numbers to be carrried with your left hand. While this is not quite the finger system Leonardo used, it works.

D. Chessboard Method

It appears the traditional method of multiplying we use today came from a method called a variety of names but most commonly referred to today as the chessboard, or scacchero, method. The name "chessboard" describes the little boxes drawn to help keep track of place value, making it look like a chessboard. As with most math methods, there were many variations to the method, including a variation with slanted lines and no little boxes at all (pictured on the next page) that appeared in the Treviso arithmetic (1478).[113]

---

112. This is the way to represent "1" in American Sign Language. In Leonardo's actual system, "1" would have been "carried" by bending the left hand small finger.
113. The line variation pictured on the next page (the one on the right) comes from *The First Printed Arithmetic*, Treviso, 1478, trans. by Smith, *Source Book in Mathematics*, 1:9. A picture from the Treviso arithmetic is also found in Smith,

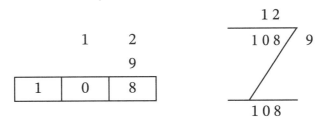

In order to better appreciate and understand this method, take a look at a larger equation: 12 × 19. Notice how the boxes or lines help make sure the correct digits are added together and remain in the correct place in the final answer. The carrying of numbers from one column to another is done mentally.

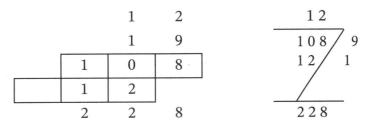

E. Gelosia Method

See page 78 for an explanation of this method.

F. Duplation Method

The duplation method reduces multiplication to a series of addition problems (multiplications by 2). We start by writing the number we want to multiply. We then double it (add it to itself, or multiply it by 2) and write the answer underneath. This process continues until we reach our answer. The left column keeps track of how many times we have added the number to itself — 24 is really 12 + 12 (12 added to itself 2 times); 48 is really 12 + 12 + 12 + 12 (12 added to itself 4 times).

1	12	
2	24	(12 + 12)
4	48	(24 + 24)
8	96	(48 + 48)
9	108	(96 + 12)

Note: The numbers shown in parenthesis are to illustrate the process.

The ancient Egyptians used this method to multiply[114] — except they used it with their writing system instead of the Hindu-Arabic decimal system we use. Since this method does not rely on place value, it can be used with a variety of different writing systems.

History of Mathematics, 2:110. General information from Smith, *History of Mathematics*, 2:107-111, and Groza, *Survey of Mathematics*, 223-224.

114. Groza, *Survey of Mathematics*, 76.

G. *Speed Mathematics Simplified* Method

While the speed method of multiplication presented by Edward Stoddard in his *Speed Mathematics Simplified* is beyond the scope of this book, I wanted to include a picture so you would be aware of it and could explore more if interested. We've described more details about the method as relates to addition on page 214.

```
   12
  × 9
  008
  108
```

H. Juan Diez Method

This method was the one given in the *Sumario Compendioso* (1556), "the earliest mathematical work to appear in the New World."[115] The work, written to help those trading in gold and silver, focused on handling specific situations traders might face more than explaining arithmetic. Along the way, though, it gives an example of an interesting multiplication method.

```
 12\9
  98
   1
 108
```

To multiply using this method, write the two numbers to be multiplied next to each other with a \ in between and a solid line underneath. Work the problem *left to right*, multiplying 1 × 9 (1 group of 10 × 9), then 2 × 9, and writing the answers below the line. The product of 1 × 9 is 9, so write a 9 in the ten's column because it represents 9 groups of 10, or 90. The product of 2 × 9 is 18, so write the 1 in the ten's column and the 8 in the one's column. Notice the 8 is written in the first available one's column rather than next to the 1. Finally, add the answers together to get the final answer of 108.

DIVISION METHODS

Below are some different division methods — have fun exploring!

Methods Covered:

A. Traditional
B. *Liber Abaci*
C. *Speed Mathematics Simplified*
D. Repeated Subtraction

A. Traditional Method

The method for division typically taught in modern textbooks is pictured on the right. See the explanation starting on page 87.

```
     19
  5)95
   -50
    45
   -45
     0
```

---

115. Smith, *Sumario Compendioso*, preface.

B. *Liber Abaci* Method

This method comes from Leonardo Pisano's famous arithmetic textbook, *Liber Abaci*. Please see the *Liber Abaci* method section on page 212 for more details about Leonardo's life and textbook.

1. Write down the number you want to divide and the number you want to divide it by, as shown.

    95
    5

2. Five goes into 90 ten time, with a remainder of 40.

    4 — Remainder: notice we have put the 4 in the ten's place to represent 40.
    95
    5 — Number of times 5 goes into 90: notice we have put the 1 in the ten's place because it represents 10.
    1

3. Mentally connect the 4 sets of 10 we have as a remainder with the 5 in 95, giving us 45. 5 goes into 45 9 times, with a remainder of 0.

    40 — Remainder of 0
    95
    5
    19 — 5 goes into 45 9 times

4. Final problem — the answer (19) is on the bottom.

    40
    95
    5
    19

Leonardo actually thought of things a little differently than presented here. He dealt with everything in terms of fractions. Instead of saying "5 goes into 90" or "5 goes into 9" as we might, he thought "$\frac{1}{5}$ of 9 is 1 and $\frac{4}{5}$," and therefore put a 1 below as part of the answer and a 4 up above as the remainder; he gave a table of common fractions for people to memorize, much as we memorize our division facts. He also suggested writing the number being divided into (in this case, the 5) off to the left as a part of a table as well as below the dividend, but does not appear to follow this advice on the majority of his example problems which is why it was not included in the example. Additionally, Leonardo presented a method of "division by numbers in head and hand," a predominately mental method involving forming numbers in the hand as you go.[116]

C. *Speed Mathematics Simplified* Method

This division method works similarly to the traditional method, except more of the equation is done mentally, the method works from left to right rather than right to left, and the method takes advantage of complements (numbers that add together to equal 10) to help with any necessary addition or subtraction.

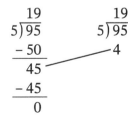

Traditional Method   *Speed Method*

---

116. See Pisano, *Liber Abaci*, 56.

Instead of writing down the 50, we mentally subtract 50 from 95 and write down the remaining 4 (note that the 4 is written in the ten's column because it represents 4 groups of 10). We then mentally connect the 4 from the ten's column with the 5 from the one's column and ask ourselves how many times 5 goes into 45.

Note: This is a very simplistic explanation of the method. There is a lot more to it. We've described more details about the method as relates to addition on page 214.

D. Repeated Subtractions

"Repeated Subtraction" is one name given to the general method used on the abacus (and by the Egyptians) to solve division problems.

While many variations of the method have been used, we will take a look at the basic method. To divide 95 by 5 using this method, we would first minus 5 from 95 to get 90. We would then subtract 5 from 90 and so forth until we could not subtract any more. The number of times we needed to subtract 5 would be our answer. (If you count the number of 5s in the figure, you will see we subtracted 5 nineteen times.) See Appendix C for an explanation of how to do this method on an abacus.

```
 95      60      30
  5       5       5
 ──      ──      ──
 90      55      25
  5       5       5
 ──      ──      ──
 85      50      20
  5       5       5
 ──      ──      ──
 80      45      15
  5       5       5
 ──      ──      ──
 75      40      10
  5       5       5
 ──      ──      ──
 70      35       5
  5       5       5
 ──      ──      ──
 65
  5
```

Bibliography

A Beka Book. *A Beka Numbers Skills K.* Pensacola, FL: Pensacola Christian College, n.d.

Adler, David A. *Roman Numerals: A Young Math Book.* New York: Thomas Y. Crowell, 1977.

Aharoni, Ron. *Arithmetic for Parents: A Book for Grownups About Children's Mathematics.* Translated from Hebrew by Danna Reisner. El Cerrito, CA: Sumizdat, 2007.

Apelman, Maja, and Julie King. *Exploring Everyday Math: Ideas for Students, Teachers, and Parents.* Portsmouth, NH: Heinmann, 1993.

Ascher, Marcia, and Robert Ascher. *Mathematics of the Incas: Code of the Quipu.* Mineola, NY: Dover Publications, 1981. Page numbers from 1997 reprint.

Bartlett, James. "How Did Jesus Study Algebra? Toward Dispelling the Fear, Dislike, and Secularization of Algebra and Advanced Mathematics." Special Report. http://www.biblicalconcourse.com/jesusalgebra.pdf (accessed April 5, 2009 - also available from a link off the http://www.biblicalconcourse.com bookstore).

Calvert School. *Calvert Math, Grade 7.* NP: Calvert School, 2001.

Cajori, Florian. *A History of Mathematical Notations: Two Volumes Bound as One.* Mineola, NY: Dover Publications, 1993. First published in 1928 and 1929 by Open Court Publishing. Citations are to this Dover Two Volumes Bound As One edition.

———. *A History of Mathematical Notations.* Vol. 1. *Notations in Elementary Mathematics.* La Salle, IL: The Open Court Publishing, 1928.

———. *A History of Mathematical Notations.* Vol. 2. *Notations Mainly in Higher Mathematics.* Chicago: The Open Court Publishing, 1952.

Clemson, Wendy, David Clemson, and Chris Perry. *Using Math...* Series. Milwaukee, WI: Gareth Stevens Publishing, 2005.

Consortium for Mathematics and Its Applications. *For All Practical Purposes: Mathematical Literacy in Today's World.* Video on Demand. Found on http://www.learner.org. Produced in 1988. (No longer available online.)

Cummins, Sereta A. *Horizons Mathematics 1: Teacher Handbook.* Part 2. Chandler, AZ: Alpha Omega Publications, 1991.

Dyson, Freeman J. "Mathematics in the Physical Sciences." In *The Mathematical Sciences: A Collection of Essays.* Cambridge, MA: Massachusetts Institute of Technology, 1969.

Euclid. *The Thirteen Books of Euclid's Elements Translated from the Text of Heiberg with Introduction and Commentary by Sir Thomas L. Heath.* 2nd ed. Rev. with additions. Vol. 1. New York: Dover Publications, 1956.

Gardner, Robert. *Science Projects About Math.* Berkely Heights, NJ: Enslow Publishers, 1999.

Groza, Vivian Shaw. *A Survey of Mathematics: Elementary Concepts and Their Historical Development.* New York: Holt, Rinehart, and Winston, 1968.

Guedj, Denis. *Numbers: The Universal Language.* New York: Harry N. Abrams, 1996.

Gullberg, Jan. *Mathematics: From the Birth of Numbers.* New York: W. W. Norton, 1997.

Hacking, Ian. *The Emergence of Probability: A Philosophical Study of Early Ideas about Probability, Induction and Statistical Inference.* New York: Cambridge University Press, 1975.

Hake, Stephen, and John Saxon. *Math 54: An Incremental Development.* 2nd ed. Norman, OK: Saxon Publishers, 2001.

———. *Math 65: An Incremental Development.* 2nd ed. Norman, OK: Saxon Publishers, 1995.

———. *Math 76: An Incremental Development*. 2nd ed. Norman, OK: Saxon Publishers, 1992.

Hawking, Stephen, ed. *On the Shoulders of Giants: The Great Works of Physics and Astronomy*. Philadelphia: Running Press Book Publishers, 2002.

Huff, Darrell. *How to Lie with Statistics*. New York: W. W. Norton, 1993.

Jacobs, Tammie D., Susan J. Lehman, Dottie A. Oberholzer, Lynnette Chevalier, Lurene Dubois, Caryn M. Moody, Debra Overly, et al. *Math 2 for Christian Schools: Teacher's Edition*. 2nd ed. Greenville, SC: Bob Jones University Press, 1993.

Jacobs, Tammie D., Susan J. Lehman, Dottie A. Oberholzer, Lynnette Chevalier, Sharon Hambrick, Caryn M. Moody, Debra Ann Overly, et al. *Math 3 for Christian Schools: Teacher's Edition*. 2nd ed. Greenville, SC: Bob Jones University Press, 1994.

Kline, Morris. *Mathematics: The Loss of Certainty*. New York: Oxford University Press, 1980.

Kogelman, Stanley, and Barbara R. Heller. *The Only Math Book You Will Ever Need*. New York: Facts on File Publications, 1986.

Labbān, Kūshyār ibn. *Principles of Hindu Reckoning: A Translation with Introduction and Notes by Martin Levey and Marvin Petruck of the Kitāb Fī Usūl Hisāb Al-Hind*. Madison and Milwaukee: The University of Wisconsin Press, 1965.

Larson, Nancy, with Linda Mathews and Dee Dee Wescoatt. *Math 1: An Incremental Development*. Home Study Teacher's ed. Norman, OK: Saxon Publishers, 2004.

Lisle, Dr. Jason. *The Ultimate Proof of Creation: Resolving the Origins Debate*. Green Forest, AR: Master Books, 2009.

Loop, Katherine A. *Beyond Numbers: A Practical Guide to Teaching Math Biblically*. Fairfax, VA: Christian Perspective, 2005.

Mattern, Joanne. *I Use Math...* Series. Milwaukee, WI: Weekly Reader® Early Learning Library, 2006.

Midonick, Henrietta O., ed. *The Treasury of Mathematics: A Collection of Source Material in Mathematics Edited and Presented with Introductory Biographical and Historical Sketches*. New York: Philosophical Library, 1965.

Nickel, James D. *Mathematics: Is God Silent?* Rev. ed. Vallecito, CA: Ross House Books, 2001.

Nickel, James D. *Rudiments of Arithmetic: Foundational Principles in the Computation and Theory of Numbers*. 1st ed. (preliminary draft). U.S.: James D. Nickel, 2008.

Pappas, Theoni. *More Joy of Mathematics: Exploring Mathematics All Around You*. 8th printing. San Carlos, CA: Wide World Publishing/Tetra, 1998.

———. *The Joy of Mathematics: Discovering Mathematics All Around You*. Rev. ed. San Carlos, CA: Wide World Publishing/Tetra, 1998.

Pisano, Leonardo. *Fibonacci's Liber Abaci: A Translation into Modern English of Leonardo Pisano's Book of Calculation*. Translated by L.E. Sigler. New York: Springer-Verlag, 2002.

Poythress, Vern S. "A Biblical View of Mathematics." In *Foundations of Christian Scholarship: Essays in the Van Til Perspective*. Edited by Gary North. Vallecito, CA: Ross House Books, 1979. Available on http://www.frame-poythress.org.

Primary Mathematics Project Team. *Primary Mathematics 1A*. Textbook. U.S. ed. Singapore: Times Media Private, 2003.

———. *Primary Mathematics 1B*. Textbook. U.S. ed. Singapore: Times Media Private, 2003.

———. *Primary Mathematics 2A*. Textbook. U.S. ed. Singapore: Times Media Private, 2003.

———. *Primary Mathematics 3B*. Workbook. Part 2. Singapore: Times Media Private, 1999.

Quine, David. *Making Math Meaningful Level 2, Parent/Teacher Guide.* Richardson, TX: Cornerstone Curriculum Project, 1997.

———. *Making Math Meaningful Level 5.* Rev. ed. Richardson, TX: The Cornerstone Curriculum Project, 1997.

Rasmussen, Steve. *Key to Fractions: Adding and Subtracting.* Emeryville, CA: Key Curriculum Press, 1980.

———. *Key to Fractions: Multiplying and Dividing.* Emeryville, CA: Key Curriculum Press, 1980.

Reimer, Luetta, and Wilbert Reimer. *Mathematicians Are People, Too: Stories from the Lives of Great Mathematicians.* Palo Alto, CA: Dale Seymour Publications, 1990.

———. *Mathematicians Are People, Too: Stories from the Lives of Great Mathematicians*, Vol. 2. U.S.A.: Pearson Education, Inc./Dale Seymour Publications, 1995.

Saxon, John H., Jr. *Algebra 2: An Incremental Development.* 2nd ed. Norman, OK: Saxon Publishers, 1997.

Smith, David Eugene. *A Source Book in Mathematics.* Vol. 1. *General Survey of the History of Elementary Mathematics.* New York: Dover Publications, 1959.

———. *History of Mathematics.* Vol. 2. *Special Topics of Elementary Mathematics.* New York: Dover Publications, 1958.

———. *The Sumario Compendioso of Brother Juan Diez: The Earliest Mathematical Work of the New World.* Boston and London: Ginn, 1921.

Stoddard, Edward. *Speed Mathematics Simplified.* Dover ed. Mineola, NY: Dover Publications, 1994.

Tiner, John Hudson. *Exploring the World of Mathematics: From Ancient Record Keeping to the Latest Advances in Computers.* Green Forest, AR: Master Books, 2004.

Vorderman, Carol. *How Math Works.* London: Dorling Kindersley, 1996.

Wilson, Robert. *Astronomy Through the Ages: The Story of the Human Attempt to Understand the Universe.* Princeton, NJ: Princeton University Press, 1997.

Wright, Michael, and M.N. Patel, editors. *How Things Work Today.* New York: Crown Publishers, 2000.

Zimmerman, Larry L. *Truth & the Transcendent: The Origin, Nature, & Purpose of Mathematics.* Florence, KY: Answers in Genesis, 2000. Also available at https://answersingenesis.org/answers/books/truth-transcendent.

Index

algebra 37–39. *See also* exponents
abacus 199–209
    addition and subtraction 56–57, 205–206
    construction of 201–204
    counting table 56–57, 200, 203–209
    decimals 131–132, 202–203
    division 87, 208–209
    history and use 55–57
    multiplication 207
    Roman numerals 35, 204
    sets 202
    soroban & suan pan 199–200
        Speed Method based on 214
Adam naming the animals 19
addition
    on an abacus 56–57, 205
    of decimals 127–128
    of fractions 103–104, 109, 118–119
    foundational concept 43–48
    methods 55–58, 212–215
    multi-digit operations 55–62
    plus sign 45
algorithms 56, 211–220. *See also each individual concept*
animals, naming of 19
applications. *See also individual concepts*
    importance 14–15
approximately equal (≈). *See* rounding
architecture 143–144, 146–147
arithmetic, concepts in 17, 165
art 110, 133, 143–144, 146
associative property 65–66
astronomy 26–27, 81–83
attributes of God. *See* God, attributes of
Babylonian numerals
    accomplishments 32
    fractions 95
    number system 197
Bhāskara method (addition) 214–215
biblical accounts referenced
    Adam naming the animals 19
    genealogies 29
    miracles 44–45
    Nebuchadnezzar 126

    Tower of Babel 25–26
    woman pouring perfume 86
binary numbers 35, 197
blood vessels 183
body (human) 19, 22, 183–184
bones, Napier's. *See* Napeir's rods
business
    addition and subtraction used in 61
    decimals used in 133–134
    ratios used in 146–147
    starting a business 61, 92
    worksheets 176, 179–180
calculus 165. *See also* algebra; exponents
calendar 29
carbon-14 dating 163
categorization 153
check writing/checkbooks 54, 60, 132
chessboard method (multiplication) 216–217
Chinese abacus. *See* abacus, suan pan; *See also* Sangi boards
commutative property 65–66
comparing numbers 37–41
complements. *See Speed Mathematics Simplified*
complex numbers 149–150
computers
    binary and hexadecimal numbers 35, 197
    cryptology 151
    graphics on 133
    written numbers, example of 187
conventions 72–73
cooking
    counting 18
    fractions 99, 109–110, 111, 112
    ratios 139, 146
counting 17–23
    by groups 151, 202
counting numbers 149–150, 162
counting table. *See under* abacus
cross multiplication 140–142
cryptology 151
currencies (monetary) 145–146
decimal point 126–127

decimals 125–134. *See also* fractions; percents
    adding 127–128
    converting from fractions 129
    decimal point 126–127
    dividing 129
    foundational concept 125–127
    multiplying 128
    on an abacus 131–132, 202–203
    subtracting 127–128
    worksheets 181–183

degrees 122

denominator 95–96. *See also* least common denominator

devices (historic). *See* abacus; Napier's rods; quipu; Sangi boards

difference 53

distributive property 65–66

dividend 71–72, 73

division. *See also* fractions; ratios; roots
    on an abacus 87, 208–209
    of decimals 129
    of fractions 103, 107–108, 122–124
    foundational concept 71–76
    fractions express 93, 99, 122–124
    long 87–92
    methods 87–90, 218–220
    multi-digit operations 87–92
    ratios express 139–140
    seeing if divisible by 115
    symbols 72
    terms 71–72, 73
    ways to write 74
    worksheets 175, 179–180

division sign 72

divisor 71–73

duplation method (multiplication) 217

dust board method (multiplication) 215–216

Egyptian numerals and methods
    accomplishments with 32
    division 220
    fractions 95
    multiplication 217
    numerals 196
    plus sign 45

Einstein, Albert 50–51

electrons 159, 161–162

equal sign 37–39

equivalent fractions 96. *See also* fractions, reducing/simplifying; factoring

even numbers 149–150

evolution. *See* exponents

exponential growth and decay 163–164

exponents 157–235
    exponential growth and decay 163
    Greeks' view of 158–159
    Pythagorean theorem 162–163, 186
    scientific notation 159, 161–162
    worksheet 186

extremes (in ratios) 140–142

factoring 115–124
    common factor 116–118
    factor trees 116–118, 120–121
    greatest common factor 116–118, 122
    least common denominator 119, 122
    least common multiple 119
    lowest common denominator 119, 122
    lowest common multiple 119
    prime factors 116

Fibonacci, Leonardo. *See Liber Abaci*

Fibonacci sequence 143–144, 151, 154, 185. *See also* sequences

finger reckoning 55. *See also Liber Abaci*

fractions. *See also* decimals; division; factoring; percents; ratios
    adding 103–105, 109–113, 118–119
    decimals, converting between 129
    different ways to write 95–96
    dividing 103–104, 107–114
    division, fractions express 93, 99
    equivalent 96–98. *See also* fractions, reducing/simplifying
    exponential 157
    factoring, GCF, and LCM/LCD 115–124
    foundational concept 93–101
    improper 96–97
    mixed numbers 96–97
    multiplying 103, 105–107, 108–114
    operations 103–114
    proper 96–97
    reciprocal, multiplying by 140–141
    reducing/simplifying 96, 100, 113, 122–123. *See also* fractions, equivalent; factoring
    subtracting 103–105, 108–114, 118–119
    terms 95–96

games 23, 48, 134, 137

gelosia method (multiplication) 78–79

genealogies 29

geometry. *See also* measuring; Pythagorean theorem
   Egyptians used 32
   exponents in 158–159, 161–163
   ratios and proportions in 142–144
God, attributes of
   able to do all He says 61–62, 74
   all-present 52–54
   in charge 52–53
   the Creator 17
   caring 22–23, 82, 143–144, 147
   consistent 27
   creative 138
   decider of truth 65–66, 167
   faithful 43–44, 47
   great 22–23, 82
   infinite 151
   loving 19–20, 82
   powerful 70, 74, 76, 92
   the Savior 19–20, 167–168
   the Sustainer 32, 73–74, 129
   source of wisdom 37, 41
golden mean. *See* golden ratio
golden ratio 143–144, 151, 154. *See also* golden ratio
golden rectangle 143–144
graphing 136, 138
greater than sign 37–39
Greek numerals and views
   exponents, view of 158–159
   fractions 95
   plus sign 45
growth 184. *See also* exponential growth and decay
hands 19, 22
Hebrew numerals 196
hexadecimal numbers 35, 197
hieroglyphics. *See* Egyptian numerals and methods
history. *See also individual concepts*
   importance of 14
   integrating with math 29–30
   of modern math 55–56. *See also Liber Abaci*
   overview of origins (Genesis) 17–20, 25–26
humanism 13
idea notebook 23
imaginary numbers 149–150
Incan numerals. *See under* number systems
infinite numbers 151
integers 149–150
interest (on money) 136, 138, 160
involution. *See* roots

irrational numbers 149–150, 158–159, 162
Japanese abacus. *See* abacus, soroban
Juan Diez method (multiplication - *Sumario Compendioso*) 218
Kepler, Johanes 113
laws, mathematical 43–44
   falsely attributed to man 50–51, 158–159
   properties 65–66
least common denominator 119, 122. *See also* factoring
least common multiple 119. *See also* factoring
less than sign 37–39
Liber Abaci
   addition and subtraction 212–213
   division 219
   multiplication 216
light (speed of) 26–27
light-years 81–83
lines, number 41
logarithms 69
lowest common multiple 119. *See also* factoring
lowest common denominator 119, 122. *See also* factoring
maps 132, 146
mathematicians 193–194
means (in ratios) 140–142
measuring
   around the house 29, 41, 111
   in card making 110–111
   to compare 41
   height of a tree 142–143
   metric 145–146
   with negative numbers 151–152
   proportions in body 184
   unit conversion 145–146
methods, math 211–220. *See also individual concepts;* abacus; Napier's rods; quipu; Sangi boards
minuend 53
miracles 44–45
mixed numbers 96–97
money. *See also* business; check writing/checkbooks; interest (on money); work
   applications with 60, 67, 75, 94, 112, 120, 123, 179–180, 181–182, 187
   bills 75, 120
   counting 151, 187
   currency conversion 145–146

decimals with 130, 132–133
perspective on 86, 132
motion 123, 152
multiple 115
multiplication. *See also* exponents; skip counting; cross multiplication; factoring
    of fractions 103, 105–108, 114
    on an abacus 207
    foundational concept 63–70
    methods 77–81, 215–218. *See also* Napier's rods
    multi-digit operations 77–86
    multiplication table 63–65
    properties of 65–66
    worksheets 173–174, 176–178
multiplication table 63–65
music 94–95, 133, 151, 154, 171
naming the animals 19
Napier, John 69. *See also* Napier's rods
Napier's rods
    instructions 68–69
    overview 64–65
    template 173
    worksheets 174–175
naturalism 13
natural numbers. *See* counting numbers
Nebuchadnezzar 126
negative numbers 149–150, 151–152, 158–159
Newton, Sir Isaac 113
non-integers 149–150, 162
non-real numbers 149–150
number lines 41
numbers. *See also* number systems; sequences
    greater than/less than/equals 37–39
    patterns 151, 154, 202. *See also* Fibonacci sequence
    types of (number sets) 149–155, 202–203
    written 25–30, 171–172
number systems 195–198
    Aztec 195
    Babylonian 32, 95, 197
    binary/hexadecimal 35, 197
    Egyptian 32, 45, 95, 196, 217, 220
    of fractions 95–96
    Greek 45, 95
    Hebrew 196
    Incan 32, 34–35, 197–198
    Roman 35, 96, 198, 204
    ways to write "one" 26

numerals. *See* numbers; number systems
numerator 95–96
odd numbers 149–150
order of operations 72–73
patterns. *See under* numbers
percents 135–138. *See also* decimals; fractions
perfume, woman pouring 86
physics 124, 152, 163
Pisano, Leonardo "Fibonacci". *See Liber Abaci*
place value 31–36, 126, 197–198
plants 22
plus sign 45
positive numbers 149–150
prime factors 116. *See also* factoring
prime numbers 116, 150
probability 155
properties 65–66
proportions 139–147. *See also* golden ratio
    cross multiplication 140–142
    worksheets 184–185
Pythagoras 162. *See also* Pythagorean theorem
Pythagorean theorem 162, 186
quipu
    accomplishments with 32
    description of 197–198
    instructions for making 34–35
quotient 71–72, 73
raindrops 44–45
rational numbers 149–150
ratios 139–147. *See also* proportions; golden ratio
    cross multiplication 140–142
    division, ratios express 139–140
    worksheets 184–185
real-life applications. *See* applications
real numbers 149–150
reciprocal (of fractions - multiplying by) 107, 140–141
rectangle, golden 143–144. *See also* golden ratio
reduction (of fractions). *See* fractions, reducing/simplifying
repeated subtraction method (division) 208–209, 220
rods, Napier's. *See* Napier's rods
Roman numerals 35, 96, 198, 204
Rømer, Ole Christenson 27

roots 157–163
    exponential growth and decay 163
    Greeks' view of 158–159
    Pythagorean theorem 162, 186
    worksheet 186
rounding
    teaching importance of 112
    examples of approximating 81–82, 111–112, 143, 179–180, 183, 184–186
rules 14. *See also individual concepts*
Sangi boards 55
science 133. *See also* Fibonacci sequence; laws, mathematical
    astronomy 26–27, 81–83
    carbon-14 dating 163
    categorization 153
    electrons 159, 161–162
    exponential growth and decay 163
    golden ratio 143–144, 151, 154
    human body 19, 22, 183–184
    integrating with math 14
    light (speed of) 26–27
    light-years 81–83
    mathematical relationships 39
    motion 123, 152, 163
    physics 123, 152, 163
    plants 22
    raindrops 44–45
    recording small numbers 159, 161–162
    scientific notation 159, 161–162
    simple counting ideas 22
    sunflowers 185
    temperature (recording) 35, 151, 187
    worldview in 11–12
scientific notation 159, 161–162
sequences 151, 154, 202. *See also* Fibonacci sequence
sets. *See also* sequences; skip counting
    basic, abacus with 202
    notation 150, 152
    number sets 149–155
set theory 155
shopping
    addition 60
    decimals 125, 132
    division 91
    fractions 100, 104, 111–112
    multiplication 68, 82–83
    percents 138
    pretend store 133, 138
    subtraction 60
    worksheets 171–172

simplifying fractions. *See* fractions, reducing/simplifying
skip counting 151, 154, 202
soroban. *See under* abacus
Speed Mathematics Simplified
    addition and subtraction 214
    division 219–220
    multiplication 218
sports 100, 133, 145
square roots 157, 162. *See also* roots; Pythagorean theorem
stars. *See* astronomy
statistics
    misused 136
    probability 155
    in sports 133
suan pan. *See under* abacus
subtraction
    on an abacus 56–57, 206
    of decimals 127–128
    of fractions 103–105
    foundational concept 49–54
    methods 55–57, 212–215
    multi-digit operations 55–62
    terms 53
subtraction, repeated. *See* repeated subtraction
subtrahend 53
sunflowers 185
symbols 37–39, 41. *See also individual concepts*
teaching. *See individual concepts*
    main principles 14–15
telling time 29, 121–122
temperature, recording 152
Tower of Babel 25–26
uses (of math). *See* applications
whole numbers. *See* counting numbers
work 18, 134. *See also* business; money
worksheets 169–186
worldview, overview of 11–15
written numbers 25–30, 171–172. *See also* number systems

ABOUT THE AUTHOR

For more than a decade, Katherine has been researching, writing, and speaking on math, along with other topics. Her books on math and a biblical worldview have been used by various Christian colleges, homeschool groups, and individuals, and her junior high math curriculum is changing how students see math. Her short stories (found on www.ChristianPerspective.net) have reached around the world. Her favorite thing to do is to talk with people about God and His Word. After all, all our biographies will end with seeing Him face to face.

When You've Finished Arithmetic...

Wondering what to do with your junior high and high schooler? Katherine has developed a junior high curriculum that teaches math from a biblical worldview (*Principles of Mathematics: Biblical Worldview Curriculum*), a video eCourse to go along with *Jacob's Elementary Algebra* (and one is due later this year for *Jacob's Geometry*), and an Algebra 2 program (*Principles of Algebra 2: Applied Algebra from a Biblical Worldview*). You can find all the books at MasterBooks.com and the eCourse supplements at MasterBooksAcademy.com.

> **"THANKS TO MASTER BOOKS, OUR YEAR IS GOING SO SMOOTHLY!"**
> — SHAINA

Made for "Real World" Homeschooling

FAITH-BUILDING

We ensure that a biblical worldview is integral to all of our curriculum. We start with the Bible as our standard and build our courses from there. We strive to demonstrate biblical teachings and truth in all subjects.

TRUSTED

We've been publishing quality Christian books for over 40 years. We publish best-selling Christian authors like Henry Morris, Ken Ham, and Ray Comfort.

EFFECTIVE

We use experienced educators to create our curriculum for real-world use. We don't just teach knowledge by itself. We also teach how to apply and use that knowledge.

ENGAGING

We make our curriculum fun and inspire a joy for learning. We go beyond rote memorization by emphasizing hands-on activities and real-world application.

PRACTICAL

We design our curriculum to be so easy that you can open the box and start homeschooling. We provide easy-to-use schedules and pre-planned lessons that make education easy for busy homeschooling families.

FLEXIBLE

We create our material to be readily adaptable to any homeschool program. We know that one size does not fit all and that homeschooling requires materials that can be customized for your family's wants and needs.

VISIT **MASTERBOOKS.COM** — *Where Faith Grows!* — TO SEE OUR FULL LINE OF FAITH-BUILDING CURRICULUM OR CALL 800-999-3777.